MW00966846

Going Places

Pauline Moser Longest

Going Places

Best wishes to Mary Helen

Pauline

PAULINE
MOSER
LONGEST

AUTHOR & ILLUSTRATOR

PENTLAND PRESS, INC.

ENGLAND·USA·SCOTLAND

PUBLISHED BY PENTLAND PRESS, INC.
5124 Bur Oak Circle, Raleigh, North Carolina 27612
United States of America
919-782-0281

ISBN 1-57197-047-9
Library of Congress Catalog Card Number 96-71763

Printed in the United States of America

Table of Contents

Europe *(continued)*

Foreword

Several of my friends have suggested that I write about my travels, and now that I have entered the twentieth century and have a computer, I am getting started on what I hope will be a reasonably interesting record of my going places. I know that everyone who suggested that I undertake this project was thinking about my trips to faraway places, but since I was born in 1913 and remember when travel was quite different, I plan to begin with some of my earlier recollections. This will not, however, be a story of my life, and when there are references to my husband, family, friends, school or teaching, or other parts of my personal life, it is because they are related to my travels.

Appreciation

To Phebe Emmons and Georgia Mullen, and others who have been tolerant of my shortcomings and have been such wonderful traveling companions.

To all the tour companies and travel agencies and their representatives who have planned and served as directors and guides.

To Gwen Simmons for being confident that I could learn to use a computer and word-processing and for her untiring patience in teaching me and helping me all along with this story.

To Marvin O. (Smitty) Smith for looking after my house when I was away. Without his devoted attention, I could not have left home with confidence that everything would be cared for; and to his wife, Helen, for the food I often found in my refrigerator when I returned home.

To my husband, and especially to my father, both of whom worked hard and invested well, so that I have had the means for fulfilling my dreams of seeing some of the world. A teacher's salary and the retirement payments would not have enabled me to get very far from home.

To whatever genes and good fortune have enabled me to maintain excellent health and the stamina which has sometimes been required, I know that I have been truly blessed.

Part I
Early Memories of My Travels

Until High School Graduation, 1929

Our family lived in Winston-Salem until I was almost seven years old, and the earliest memory I have of a motorized vehicle is the streetcar which ran on our street.

Winston-Salem Streetcar —1918
Drawn from memory.

I don't recall much about riding in the car, but I believe that sometimes when some of us children were playing in the street the conductor would let us ride to the end of the line—about two blocks. Playing in the street was acceptable then, because the traffic consisted almost entirely of horse-drawn carts of the milkman or the iceman. The thing I remember best about the streetcar was that someone had discovered that if two straight pins are crossed and placed carefully on the track the weight of the car would flatten them into something resembling an open pair of scissors. Incidentally, don't try this with modern pins with colored plastic heads—ours were all metal.

The first automobile I remember seeing was about a block away, and most of the folks in the neighborhood were there looking at it. I think it was what was called a Copperhead Ford, because pictures I have seen of that model in later years had the same brassy-looking heading as the one I remembered. The only airplanes we saw during those days were some distance away in the sky doing stunts. When one of those was spotted, word spread and everybody came out to look. I can still picture the loop-the-loops of one of those small planes.

When we went out of town it was to visit relatives in the country—about ten miles away. I remember Mama's father coming for us in a horse-drawn carriage. Later on, Daddy's brother bought a Dodge automobile, and we had the experience of riding in that. Actually it wasn't so great, because the pavement gave out at the city limits and the country roads were rough. Once, I remember—it must have been on our Christmas visit because it was very cold—the Dodge got stuck in the mud. We were stalled for a good while until a man who lived nearby could be called to hitch up his mules and pull us out.

In 1920 my dad bought a Chevrolet so he would have transportation to work, and we moved to the country. The road to Lewisville from Winston-Salem had been improved, and some traveling road builders were preparing to improve the road through the village. The procedure was to make what was called a "sand-clay" road. First the road was widened and graded, then a coat of red clay was applied. This was followed by a covering of topsoil, much to the distress of the landowners who had to supply the soil. The labor was done by manpower or by mule-drawn implements. While this was a great improvement, the road was dusty in dry weather and subject to weathering when it rained. Since all of the roads did not get even this much improvement, getting stuck in the mud remained a common occurrence.

Another hazard for drivers was the flat tire, and knowing how to change and repair a tire was as important as knowing how to drive. The tire had to be taken from the wheel and the inner tube removed. If there had been a puncture, it could possibly be repaired. Every sensible driver carried a patching kit, as well as a spare tire. The tire was then put back on the wheel, and this was not an easy job. Then air was pumped into the tire with a hand pump.

Fortunately our family lived a short distance from the school, because some of the pupils had to walk several miles. Three sisters who lived some distance away came by horse and buggy. I liked to watch them undo the harness in the morning and hitch up again in the afternoon.

During my years in grade school roads improved somewhat and the number of cars increased. If we had to go to town (Winston-Salem) to shop we usually had to go on a Saturday since Daddy had a load of passengers on weekdays. This was not the same as car-pooling—these people usually did not have a car. On the Fourth of July we often packed a lunch and took a ride to some place of interest. I remember going to Moore's Springs, Pilot Mountain, and Mount Airy, where I was greatly impressed by the large granite quarries.

About 1923 I saw my first school bus. A small section of Yadkin County had been added to Forsyth, and this bus was provided to bring the children from that area to Lewisville School. The bus was small, painted black, and the body was very high off the ground. For several years after, buses became more and more common, as a number of one- or two-teacher schools were consolidated into Lewisville.

During my later high school years some of the boys had cars of a sort, or could borrow the family car at times. Now and then some of us got together for a Saturday night movie or to ride around the countryside on Sunday afternoon. We took lots of pictures with a Brownie camera.

Until College Graduation, 1933

In the fall of 1929 I had my first train ride, from Winston-Salem to Greensboro, and began my four years at what was then known as the North Carolina College for Women. I became quite familiar with bus and train schedules, as I didn't do much other traveling during those years. One exception is that as a member of the Botany Club I went for three years on their camping trip to the area now included in Hanging Rock State Park. One sight always comes to mind when I think of those trips. On one occasion, as we were walking away from the truck which had brought our supplies, I looked back and saw a complete rainbow. It arched over the sky between two small mountains and then came down in front of the mountains.

Each year the students on the trip, along with a couple of professors, spent one day making a hike to the top of Moore's Knob, the highest peak in the Sauratown Mountains. Of course we learned some botany along the way, but the amazing view from the top of the mountain impressed me more than anything else. At night we would go down a very treacherous trail to the bottom of a waterfall we knew as Cascade Falls. The object was to go behind the falls and see the glowworms on the rock wall.

I should also mention in connection with the botany group that in the spring of 1932 my laboratory partner and I were determined to collect and identify the two hundred species of flowering plants which our professor had set as the goal for our local flora class. We spent each

Saturday, weather and other conditions permitting, walking to Guilford College—about seven miles away. We carried our collecting paraphernalia and gathered our specimens as the blossoming season progressed. We not only walked the distance, but we crisscrossed the road and went into available open fields or forests some distance off the road. When I think of that now I wonder what would happen to two teenage girls who would undertake such a venture. Yes, we had 202 specimens at the end of the course; most of these were what many people would call weeds.

✄   ✠   ✄   ✠   ✄

During the summer of 1932 I saw the Atlantic Ocean for the first time. Our favorite minister had been sent to Avon to look after the church there and he invited some of our Epworth League group to visit him and his family. Nine of us accepted, and on a Sunday afternoon we set out in two cars for the Outer Banks. At that time the only scheduled way to get to Avon was by means of the mail boat from Manteo. I believe that it made two round trips each week. We had to go all the way to Point Harbor and Sligo to get a bridge to the

Cape Hatteras Lighthouse

Banks, drive by Kitty Hawk, Kill Devil Hills, and Nags Head and then across the Manteo bridge. We reached Manteo about three o'clock A.M., and the boat left at six. We tried to sleep on the grounds of the Dare County courthouse, but the mosquitoes would not cooperate.

Finally we left the cars in Manteo and were on our way to Avon.

At that time a number of the inhabitants had never left the island, and we had difficulty understanding their speech. There were a few cars, which did not have to have a license tag because there were no roads. We went fishing in the Pamlico Sound almost every morning and had to walk through hordes of mosquitoes to get to the pier. This was long before women started to wear pants, but in order to protect our legs from the mosquitoes we borrowed pants from the men. This, unfortunately, resulted in cancellation of our expected deep-sea fishing trip. The owner of the boat which was to take us saw us coming from our morning expeditions and said that his religious beliefs would not permit him to take on his boat women who wore men's clothes. To compensate for our disappointment we took a day to walk to Cape Hatteras–about seven miles on the tide wash. When we got there we climbed to the top of the lighthouse. Members of the Coast Guard stationed there invited us for lunch, and then drove some of the less sturdy ones back to Avon. I walked back.

Until Marriage, 1941

On 26 August 1933, my twentieth birthday, a couple of friends drove me to Murphy to begin my teaching career. A year or so ago I heard someone say that she did not like to drive from Asheville to Murphy because the road curved so much. I took that ride not long ago and marveled at how straight it is. You had to have gone that way as it was in 1933 in order to know what a real curving road can be. I'm talking major hairpin curves, one after the other. It was scary, but I am glad that I had the opportunity to see the magnificent views from the bottom of the Nantahala Gorge.

In those days the only teachers who had cars were coaches, couples who were both working or those who had another job at night. (This was called moonlighting and was frowned upon by superintendents and school boards.) Others had help from their families. My salary was $560—I mean per year—$70 per month for eight months. A man who began teaching the same year I did told me recently that the price of a new Ford car in 1933 was $560. This was the basic state salary; some of the cities and well-to-do counties paid a supplement.

A few of the Murphy teachers who had cars were generous in including the rest of us on their trips in and out of town. One Saturday a group of us went to the area of Ducktown, Isabella and Copper Hill, Tennessee. Copper mining had been practiced there for some time, and the gases from the ore treatment had been released into the air. This had caused all of the vegetation in the area to die, and the land was a series of bare red hills and gullies and an occasional muddy stream. We were told that the air had been cleaned, and that efforts were being made to bring back the vegetation, but did not see that any progress had been made.

Some members of the group who had been there before were anxious for the rest of us to see a school building in Isabella. It was a beautiful brick building, and a caretaker let us go inside. The furnishings and supplies and the construction itself were so attractive that we, in this time of depression, were amazed. After a while some-one asked how long it had been used, because it looked so new and clean. The answer was that even though it had been there for several years it had never been used. We were told that some time ago the school board had purchased stock in a copper mine, and this building had been built with returns from the investment. Then an election had taken place. A new school board had been elected, and had refused to use the building. A little wooden structure not far away was pointed out to us as the school in use.

Sometime along in March there was an epidemic of measles in the Murphy school. Many of the pupils had the disease and the parents of others were afraid to send them to school. Consequently the school was closed for three weeks. After we heard this my roommate (from Salisbury) and I decided to go home by train. Since we didn't have anything else to do during the ride, we kept a list of all the stops made by the train from Murphy to Asheville. Although the distance was about ninety miles, we recorded thirty little stations. The train the rest of the way home was not so slow.

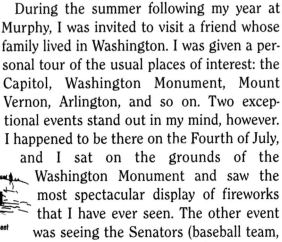
Washington Monument

During the summer following my year at Murphy, I was invited to visit a friend whose family lived in Washington. I was given a personal tour of the usual places of interest: the Capitol, Washington Monument, Mount Vernon, Arlington, and so on. Two exceptional events stand out in my mind, however. I happened to be there on the Fourth of July, and I sat on the grounds of the Washington Monument and saw the most spectacular display of fireworks that I have ever seen. The other event was seeing the Senators (baseball team, that is) play the New York Yankees. Our seats were near the front on the first base line, and I got a good look at Lou Gehrig.

The next year I went to Onslow County to teach. The first year was in Dixon and the next six in Richlands. During those years I became much better acquainted than I wished to be with Carolina Trailways, which was usually my transportation to and from home.

In late February of my first year at Richlands we had a heavy snowfall, and school was closed. Although the snow melted within a few days, roads were so muddy that the buses could not run. After a few more days we were told that school would resume the next day. It snowed again that night, so we went through another week of vacation. Unbelievably, this happened a third time, and we had to make up three weeks of school. Governor Kerr Scott, with his program for paving all roads which were traveled by school buses, came several years later.

During the summer of 1936 I had my first long trip. The mother of one of my students was planning to drive to California for her brother's wedding, and wanted someone to go along and help pay expenses. I recruited two friends, and we set out for the West Coast. There was not much excitement until we got to Dallas. Texas was celebrating the one-hundredth anniversary of its freedom from Mexico by a big exhibition in Dallas. I had never seen anything like this before, and I thoroughly enjoyed the displays and the entertainment. We went from there to El Paso and took time to cross the border and see some of the sights in Juarez.

This trip was before the days of motels. While the cities most certainly had hotels, we were on such limited budgets that we spent most of our nights in homes which advertised overnight facilities for tourists. These places were not always easy to find, but we managed all right.

Our driver had told us that we must time ourselves so that we would cross the Arizona and California desert at night, but she got in too big a hurry to get to her destination and that plan did not work out. We spent a night in Flagstaff, Arizona, which I remember as a very beautiful place because it had something we had been missing for several days–trees. Next morning we got up for an early start to the Grand Canyon, but didn't get to see much of it before we were hurried on our way. We reached the area of Needles just in time for the hottest part of the day, which was HOT.

For several miles we rode with the car windows closed, because the air which blew in when they were open was so hot it burned us. Obviously the heat in a closed car was not fatal, because we did survive, even though there were times when we had doubts. After the sun went down we stopped at a little town for food and gas and asked what the temperature had been that day. It was 116! I believe that even if a car air conditioner had been available this would have been too much of a challenge for it.

After we arrived in Fullerton we spent a day or so with the family of our driver, and then she drove us to Los Angeles. We found an inexpensive but comfortable hotel and spent almost two weeks visiting sites in and around Los Angeles. We became familiar with the local trains and buses and saw lots of missions, Forest Lawn Cemetery, the Rose Bowl, Huntington Art Gallery, Riverside, Santa Monica, Long Beach, Hollywood, and so on. We even took the overnight cruise to Catalina.

As we were waiting for a bus one Sunday, the Fourth of July, we spoke to some people who were on their way to hear Aimee Semple McPherson. We decided to go along. Angelus Temple was a monstrous building with a domed ceiling, a full stage and orchestra area, and seating for mobs of people. Part of the program was entitled "Parade of the Flags." This was a series of dramatic presentations related to historical events and a particular flag. In the last scene, the Christian flag was carried by Christian to the top of a tall pole. Then the Devil appeared, complete with red suit, horns and tail, climbed to the top of the pole and tore down the flag. Christian came back, chased the Devil away, and restored the flag. At last Aimee appeared, walking down a long stairway. She was wearing a long gray gown and carrying a big bouquet of red roses. I don't remember a thing she said in her sermon, but she accompanied it now and then with a tambourine. It was quite a show!

We went back to Fullerton for the wedding, and left for home soon afterward. Our first stop was Yosemite National Park. Although we had no reservations there was no problem in getting rooms at what appeared to be the only lodge there. The next

Driving through
the Tunnel Tree,
Yosemite, 1936

day we toured the park, drove through the Tunnel Tree, and took pictures. During the years which have followed, I have seen many beautiful places in the world, but nothing has ever made as great an impression on me as Yosemite. The beautiful waterfalls, El Capitan, Half-Dome, the magnificent big trees, and the sense of wonder about it all are still vivid memories.

We missed Reno, and Las Vegas was still a small resort town known only for its healthful spring water. We spent a night in Carson City, and wondered how such an unimposing little town could be a state capital. The next day we had probably the most unremarkable drive possible as we drove through Nevada.

Interest picked up when we got to Salt Lake City. We had a guided tour of Mormon Square and a history of the Mormon religion. We rode on the salt flats and then donned our bathing suits to try out the water in which "you cannot sink." It was possible to wade out a good distance before the water got deep enough to have our feet begin to lift up, and we had been told to turn back as soon as this happened. In spite of the fresh water shower, I had salt in my ears the rest of the day.

From Salt Lake City on we were more interested in getting home than we were in looking at the scenery. One thing we had to notice, however, was the effect a locust plague had produced on the corn in Kansas. Nothing was left of the plants except a bare stalk. We had to put a screen over the front of the car to keep the locusts from clogging the grill.

We had been away from home about four weeks, and I counted my money and realized that I had spent $220. This was my share of car expenses, the overnight stops, the hotel in Los Angeles, meals, the package cruise to Catalina, some souvenirs, a wedding gift and a long dress which I had been told was necessary for the wedding. Our meals averaged less than a dollar a day. In Los Angeles we usually had breakfast at the Owl Drugstore (lots of orange juice, and eggs, toast, and coffee for fifteen cents). Lunches were about a quarter, and dinner usually not over fifty cents. Once we saw a restaurant which advertised rattlesnake steak, and I wanted to have it, but it cost a dollar. We did not go hungry, and did not lose any weight.

Until the Move to Fayetteville, 1952

I left Richlands after the close of school in 1941, and Cecil and I were married May 31. We drove to Florida soon afterward, and stayed the remainder of the summer while Cecil played baseball with the Fort Pierce Bombers. We returned to North Carolina in time for him to begin graduate work at the University of North Carolina and for me to begin teaching at Chapel Hill High School. Events of that year resulted in Cecil's going into the army and my going back to Lewisville to teach for three years. At the end of the first year Cecil was stationed at Mobile, Alabama, and I was able to go and be with him for about six weeks until he was sent overseas.

When the time came for me to go to Mobile and take the car, I had never driven except for short distances, and did not feel at all qualified to drive to Mobile. I persuaded my sister Louise to go along and drive. As we went through Opelika, Alabama, we got into a heavy rain which lasted for several miles. When we finally got out of the rain, the windshield wiper refused to be turned off, and soon got to be very annoying. In trying to cut it off, I finally broke something, and it stopped. Louise could not come back to drive me home, so Cecil gave me a few driving lessons and I started out, feeling very insecure, at four o'clock in the morning.

I got along pretty well until I got to Opelika, where it was raining again (or still) and I had no windshield wiper. I was afraid to stop, and somehow I managed to creep along and stay on the road. I got to Atlanta at about five o'clock in the afternoon, and the only way to go was through the downtown area. By then I had pretty well mastered the art of starting and stopping without jerking, but the drive through Atlanta in what was rush hour even in 1943 was almost as frightening as the rainstorm. Back to Opelika, I have since been through that town several times on the way to other places, and it has always been raining there.

After school was out in 1945, I entered graduate school at Chapel Hill, and Cecil returned from his overseas army duty on Thanksgiving Day. He taught in the high school there until I finished my degree and then we went to Indiana University where he did some graduate study. We were there for three years, and had a number of pleasant outings, often with another North Carolina couple, to beautiful state parks in Indiana and excursions to Indianapolis, Chicago, South Bend, French Lick, and so on.

During our stay in Bloomington, my mother and Louise and her daughter came to visit us. We took a couple of days to visit a friend in Jacksonville, Illinois. Having grown up on a farm, Mama was very much impressed by the black soil and the tall corn of Illinois. From Jacksonville, our friend took us to Hannibal, Missouri, to see Mark Twain's house and other sights along the Mississippi.

While in Bloomington I had the unique opportunity of becoming a member of Mrs. Kinsey's hiking group. (Yes, she was the wife of *the* Dr. Kinsey.) The invitation to join the group resulted from my having helped in the Girl Scout day camp which Mrs. Kinsey directed each summer. Each Wednesday afternoon–deterred only by heavy rain– snow on the ground made no difference–Mrs. Kinsey drove around and picked up her group of six, seven or eight and took us to an interesting wooded area for our hike. Actually it was more of a nature study than a health-promoting activity.

Until Cecil's Death, 1970

Then Cecil got a job at Fort Bragg, we moved to Fayetteville, and I started teaching at Seventy-first High School. After a year or so of car-pooling, we bought another car, and I had finally become a car owner. I became interested in the North Carolina Education Association and its classroom teacher division, and spent a good bit of time attending their various meetings. Cecil spent a lot of his spare time officiating at athletic contests and later serving as booking agent for the Fayetteville Athletic Officials Association.

Statue of Liberty

Some time during the early fifties Emmy Peek (a Fayetteville neighbor) and I took a bus trip to New York for a week. We had several city tours, and one which took us up the Hudson River, with stops at Tarrytown and Peekskill, then on to the FDR home at Hyde Park. We crossed the river at Newburgh, and had a visit to West Point on the way back. We saw several plays in the city, but the big thing for Emmy, a devoted Yankee fan, was two baseball games at Yankee Stadium. We were both especially thrilled at seeing Yogi Berra and Joe Dimaggio in action.

I attended practically every NCEA (later NCAE) state convention from the early 1950s to 1979, when I switched to the Retired School Personnel Division. There were also a number of National Education Association meetings at various times during those years. I remember especially a session at White Sulfur Springs, West Virginia. We stayed at the Greenbrier Inn, and I had never before been in such an elegant hotel. One day we had lunch at the beautiful golf club dining room, and I had things to eat that I had never heard of before.

One memory of that lunch, however, was of a different nature. One of the ladies had ordered a dessert which came piled high with whipped cream. Just as she was lifting her spoon to start eating, a big fly landed right on the top of all that cream. Needless to say, everyone except the waiters thought this was hilarious.

I went with Lois Lambie to my first National Education Association convention in 1957. It was held in Philadelphia, and we had a brief stop at Williamsburg on the way. After the convention we went to Charlottesville and had a tour of Monticello and Ash Lawn. As we were getting ready to leave Ash Lawn to go home, Lois studied her Virginia road map and decided that we did not need to go back to Charlottesville in order to get on the road to Richmond. Another friend who had traveled with Lois had said that when Lois wants to go somewhere she draws a straight line from origin to destination and goes that way whether or not there is a road.

The route she took that day demonstrated that method, except that the line was not straight. There were lots of right angle turns from one country road to another, and eventually they were not paved. The roads became more rough and narrow, until finally we were riding through the woods on a lane made only for ox-carts, and were fording little creeks. It was dusty and hot, as this was before cars came with air-conditioning. Somehow, in about twice the time it would have taken on good roads, we did reach the highway to Richmond.

✿   ✠   ✿   ✠   ✿

Cecil and I attended three of the annual reunions of the Forty-fifth Air Depot Group, with which he served in World War II. One was in South Bend, and after the meeting we went to Chicago to see the White Sox play. Another was in Columbus, and one in Cleveland. Following the one in Cleveland we went to Niagara Falls. We did the whole sightseeing bit, wearing the yellow slickers, and watched the moving colored lights on the Falls at night. That night we had a room on the Canadian side, and admired the beautifully landscaped areas. What a contrast it was to the dirty, trash-strewn United States side!

In the summer of 1960, we took a trip to New England. First we explored the Cape Cod area, and then went on to Boston, with a stop to see Plymouth Rock. Our first aim in Boston was Fenway Park, which we finally found after much wandering around the rotaries. We got to Fenway in time to see the Red Sox and the Yankees, and felt very special at seeing Mickey Mantle and Ted Williams on the same day.

I had suggested to Cecil that we would do well to take a guided tour of Boston, but he had felt sure he could drive anywhere we wanted to go. However, after the trouble we had in finding Fenway Park, he was ready to leave the driving to Greyhound. We took the all-day tour of the sights in Boston—Old North Church, Capitol, Christian Science Building, Harvard, and so on, then went to Lexington, Concord and Walden Pond. All of this was very interesting and rewarding. I must mention the stained glass model of the earth in the Christian Science Publishing House. Large enough for several people to walk inside, it gave me for the first time an understanding of the "great circle" route, which I associated with the reports of Lindbergh's famous flight.

From Boston we went on to Maine. Kennebunkport had not achieved fame by this time, so we passed it by and went on to Mt. Desert Island and Bar Harbor. We realized the appropriateness of the expression, "the rockbound coast of Maine." On across Maine to New Hampshire we enjoyed the beautiful scenery, especially the many trees which looked like perfectly shaped Christmas trees. Near Berlin, New Hampshire, we found a ski lodge with practically no occupants—it was summer—with a gorgeous view of Mt. Washington, and spent a night there. On to Vermont, where we saw more wonderful views. From then on we concentrated pretty much on getting home.

Cecil drove me to three different NEA conventions—Atlantic City in 1961, Denver in 1962, and Detroit in 1963. We did not take time to do much sightseeing on any of these trips, but I do remember especially the seemingly endless wheat fields of Kansas and Colorado. It was harvest season, and when we stopped for lunch along the way we had an opportunity to talk with some members of the harvest crew and learn something of their life as they followed the season through several states.

My first flight was to the NEA convention in Seattle in 1964. I had to learn quickly to deal with complications, because I had to change planes in Chicago. I have never had another flight since that day in which I had such a clear view of the earth below. I imagine that this was because planes did not fly at such high altitudes then as they do now. I had perfect views of farmlands, rivers and mountains. I wished so much that members of my geography class at Methodist College could have seen all of these illustrations of the formations we had been studying.

I was initiated into Delta Kappa Gamma (an honorary society of women educators) in April of 1962, and attended my first annual state convention a few weeks later. I have not missed a state convention since. International conventions are held in even-numbered years, and I attended the one in New York in 1968. With the exception of 1970, I have gone to each one since that time. Regional conferences are held in alternate years, and I have attended every one of the Southeast regional meetings from 1971 on. While these meetings are held for information and conducting of business, a good many of them have included travel for its own sake. There are also special travel and study tours and a number of other society-related expeditions which will be described later in this story.

During the sixties I was doing a good bit of hopping around to meetings, and Cecil began to try to tease me by telling friends that I never stayed at home. I could always even this up by telling how often he was away with his duties with the Athletic Officials Association. He would say that I was always going to Raleigh, Washington, New York, or whatever, but he invariably ended by adding Chocowinity—his idea of the last place anybody would want to go. Actually I had never been to Chocowinity, but one day I did have occasion to go. Phebe Emmons, a good friend since Richlands days, had become a member of the staff of NCEA, and I was a member of the board. In those capacities we were asked to conduct a meeting at Chocowinity. Phebe was to drive me home to Fayetteville and spend the night with us. It began to snow, and friends insisted that we spend the night with them nearby. However, it was March, and we should have known better, but we felt sure this was not going to be a real snow. By the time we got to Goldsboro, we had decided we had better stop overnight, but the roads were so clear that we thought we were out of the snow.

How wrong we were! By the time we got to Newton Grove (where overnight facilities would have been limited anyway) we were committed to going on. So on through Spivey's Corners we continued—the only car on the road—and the snow was getting heavier all the time. We actually would not have known where the road was except for the trees and bushes along the shoulders. Highway I-95 and the Fayetteville streets were not so bad, but we still had the Longest driveway to negotiate. In about six inches of snow, we avoided slipping off the dam and got to the house at about midnight. Cecil had waited up for us, and when he came and opened the door he simply turned away and went to bed without saying a word. We had to admit that he was justified in not trying to say what he thought.

From 1970 to 1973

I was teaching at Methodist College when Cecil died on 3 December 1970. The college was still on a schedule which ended the semester near the last of January and then had a week of vacation. I decided to take a Caribbean cruise during that time. The ship sailed from Miami, and had stops at Puerto Rico, Haiti, St. Thomas and Nassau. The Haiti stop included a ride on a small horse to the top of Cap Haitien. Two native boys were in charge of each horse, and while they played around and joked with one another in their version of French, they really knew how to handle those horses. The trail was almost straight up, and was rough and rocky, but the view was magnificent and the story of King Henri was worth hearing.

On the way, before we started the climb, I had an opportunity to see how the people lived, and could scarcely believe the degree of poverty which prevailed. The homes were small adobe structures with dirt floors, no doors, and holes for windows. Cooking was done outdoors, and clothes were washed in dirty ditches which apparently also served as sewers. The people depended largely upon sales of their carved wooden objects for income, and their workmanship was actually rather good. We were practically mobbed as they crowded around us and offered their carvings for almost any amount they could get.

This was my first experience at seeing how some of the rest of the world lives, and it furnished in part my incentive to travel as much as I have. Of course I also give credit (or blame) to reading Richard Halliburton's books while I was in high school. The friendly people on the ship and the delightful service also encouraged me to try it again.

Sometime during the winter of 1971 I received a notice from the University of Colorado that one of their professors who had spent a lot of time doing research in the Galapagos Islands was offering a course in biogeography of the Galapagos in the summer. I felt that I simply had to enroll in that, and Lois Lambie agreed to go with me. We met with the rest of the group (about thirty) in Guayaquil, Ecuador, and flew to Baltra Island. This island had been the site of an American air base during World War II and had the only airport in the islands. After we landed we had our choice between two ways of getting to the boat which was to take us to our headquarters. One was to walk, and the other was to ride in a truck which was carrying the "honey buckets" to the ocean.

The walk was very interesting in that it went through a field of dry grass infested with little white webs. Our professor told us that these were the nests of black widow spiders. We took him at his word, but wondered why they were so numerous. His explanation was that when the soldiers were stationed there during the war they had little to do, so spent their time shooting everything that moved. This meant that all birds, lizards, and other enemies of spiders had been eliminated. The spiders were probably brought in on some supplies, and reproduced freely. Thus, our first lesson in environmental imbalance.

Our boat took us to Santa Cruz, one of the only two islands which had human inhabitants. Other attempts at settling had been abandoned because of difficulty in getting water. It is, of course, futile to try to dig a well in volcanic ash or lava. Rain was the only source of fresh water, so cisterns had to be used to collect it during the rainy season. Our living quarters consisted of little cottages, with a meeting room and dining room nearby. Brackish water from the inlet was piped in for bathing and laundry, and we were furnished one bottle of fresh water each day. Since this island had been selected for the Darwin Research Station, it had electricity and other conveniences.

The time for our three-week stay had been selected because it was the end of the rainy season, but the infamous Nina fooled them and it continued to rain the first week. We had a lot of lectures, but soon had to begin our explorations. The first of these was a walk to the top of the dormant volcano which had produced the island. It was a two-day hike, with an overnight stay at a farmhouse about halfway up. Except for the truck at Baltra and a jeep at the research station, there were no motorized vehicles, so there were no roads.

We walked that day on a soggy wet trail of volcanic ash. I felt very smug because I had brought my laced-up hiking boots. Those who were wearing low shoes sometimes sank so deep that when they lifted a foot the shoe stayed in the mire and had to be fished out. We made it to the farmhouse, where there was a large cistern and strange but adequate plumbing. The story of the family there is too long to repeat fully, but they were Norwegians who had escaped from their home when Hitler's forces got near. They were quite self-sufficient, and raised pigs and chickens and a variety of fruits and vegetables.

One special memory of that night was that our host took us outdoors and pointed out the North Star. We then turned around and saw the Southern Cross. That kind of opportunity is so rare that it made up for a lot of inconvenience. Another satisfying experience was to write our names in a guest book in which the name of Thor Heyerdahl also appeared.

We studied the several strata of vegetation as we went from the ocean to the mountain-top and wondered at some of the unusual plants. While the plants were interesting wherever we went, the real attraction was the ani-

Galapagos Tortoise

mals. We went to several other islands, sometimes spending the night on the boat. We would transfer to a little dinghy, and more often than not would make a wet landing.

Many books have been written about the animals, and I will mention only a few of them. We saw the famous tortoises only a few times in their natural homes, but there were lots of them at the research station. Many large monsters were there, and also hundreds of small ones which had been hatched there and were being cared for until they grew large enough to return to the wild. The work there is of course aimed at restoring the population which was almost annihilated by various human activities.

One morning on the boat we had grapefruit for breakfast and were told to save the peels, because they were a favorite food of the iguanas we were to visit. Sure enough when we docked, crowds of them came scurrying toward us. I have a picture of one of our girls with a small three-foot long iguana sitting on her lap and eating grapefruit from her hand. A number of the great sea lions were also there that morning, but they were interested in maintaining their territory and catching fish. They did not, however, object to our patting them on the head.

We encountered some of the small penguins on one of the islands. When we tried to take pictures of them they came nearer to us, making it difficult to keep them in focus. It was as if they were curious to get a better look at us.

We saw many birds on land and overhead from the boat, but my favorite one was the blue-footed booby. His feet are not just a little bit blue, but he looks as if he had stepped in a bucket of bright blue paint. They were fairly common on several islands, but we went to one which we called the "booby hatchery." This island had a rather clear crater, and it was a nesting place for these

Blue-footed Booby
Galapagos

birds. There were hundreds of them, some still involved in courting procedures and others setting on their nests (actually bare ground). The temperature was 110 degrees, so they were sitting on the eggs not to keep them warm but to keep them from getting too hot. Mates apparently took turns with this duty, and the free one would go to catch fish to feed the other. We walked among them, but had been told not to get too close, because during breeding season they sometimes get nervous.

At the end of our stay we flew back to Guayaquil and were scheduled for a flight to Quito. Someone suggested that we take the train up the Andes instead, and we were able to make arrangements for that. The train was a special one for tourists only, and when we saw it we said, "It's a school bus." It really was, and a plaque showed that it was made in High Point, North Carolina. The wheels were different, of course, and the engine must have had increased power, but the body, seats, and driver's position looked natural. It was a steep climb, and it was often necessary to use switchbacks. Sometimes the driver had to blow the horn to get the sheep off the track, and unfortunately one of them did not move in time. The scenery was spectacular, and as we neared Quito we had a beautiful view of Mount Chimborazo, which is on the equator, but has its top covered with snow all year. We had time for a little sightseeing in Quito, and were especially impressed by the obviously very poor people who were taking gifts to the ornate cathedral.

❃　❈　❃　❈　❃

The Delta Kappa Gamma Southeast Regional Conference was held in Hot Springs, Arkansas, in 1971. A group of five of us decided to drive there, with a stop in New Orleans for several days. We were fortunate enough to get a room with a balcony overlooking Bourbon Street, and enjoyed watching the crowds. We went to several spots of interest—the cathedral, wax museum, Cabildo, Preservation Hall, Pat O'Brien's, and so on. We dined at Antoine's and breakfasted at Brennan's.

At Hot Springs we took advantage of the baths and massage and attended meetings. During one business session the travel and study committee announced that the society was planning to sponsor a trip to Russia at

Christmas time. Mozelle Causey and I looked at each other and said, "Let's go."

It was a two-week trip sponsored jointly by Delta Kappa Gamma and Kent State University. There were three hundred fifty of us and a 747 had been chartered. While it might seem foolish to go to Russia in winter, I am glad I did, because I had always pictured it as very cold. We went first to Leningrad, and although we felt pretty shaky as we went through the extensive entry procedure, we got through all right. We were then turned over to our well-trained English-speaking guides and given our bus number for the week in Leningrad.

The most impressive sight was the Hermitage, a former palace of the tsars which had been converted into a museum to house the magnificent art collections of Catherine the Great. There were paintings by Titian, Corregio, Rembrandt, Reubens, and others of fame, statues by Michelangelo, and magnificent furniture, jewelry and costumes. We attended a production of *Eugin Onegin* at the opera house. Many of the ornate buildings were painted a beautiful blue, green, or yellow, and they made a marvelous appearance in the snow.

Since we were a group of educators, we were given visits to schools and to the headquarters of the teachers organization. We heard lectures about the Russian school system, and were entertained by groups of students. Even though the Russians officially scorned any kind of religion they had a special Christmas dinner for us, with a little Christmas tree on each table.

We had a daytime train ride to Moscow, taking along a lunch of boiled eggs and sandwiches. The cheese was good, but the Russian-style sausage—coarsely ground and mostly large chunks of fat—was not palatable. The number of daylight hours was short, and it was a gloomy day anyway, but we did get an occasional glimpse of a small town or village. There were lots of pines and other evergreen trees and the white trunks of birches could be distinguished.

Buses met us at the station in Moscow and took us to the enormous Hotel Russia. Our first view from there was the brilliantly lighted St. Basil's Cathedral and the red star-shaped lights and the flags on the Kremlin towers. It was a breathtaking sight. There were several days of touring the city and visiting schools. Special events included a performance of *The Fountains* of Bakhehisarai by the Russian Ballet. We were given the privilege of breaking in a line of hundreds of Russians standing in the cold to visit Lenin's tomb. Another special event was a troika ride in the botanical gardens. Even though the view consisted of snow and barren trees, it was a thrilling ride.

A tour within the Kremlin included several churches of the Eastern Orthodox style, all of which had been made into museums. Tombs filled much of the floor space, and the walls and pillars were covered with religious paintings. The Kremlin Museum had many interesting carriages and costumes of former rulers, as well as jewelry and other treasures.

We noticed in the streets the lack of automobiles. Except for a few limousines speeding in and out of the Kremlin, traffic consisted of buses and trucks. The food was plentiful, but monotonous—lots of potato soup, some fish, very little meat, practically no fresh fruit or vegetables. A large bowl of soft-boiled eggs appeared at every breakfast. We had an impression that Russian people were decidedly not happy, because they never seemed to smile. One night about three o'clock I looked out the window and saw a solitary woman, bundled up and wearing the ever-present babushka, shoveling snow off the street.

✄　◈　✄　◈　✄

In February of 1972, my brother Richard was finishing an assignment in Puerto Rico, and his wife Clara and I flew there and planned to come home when he did. He drove us around the island, and we also took a trip to St. Thomas on one of those tiny planes. We were introduced to banana daiquiris.

In 1971, and three other times in the seventies, I went with Richard and his family to Daytona to witness the Fourth of July Firecracker 400. Thanks to having a cousin in Florida who knew the right people, we were able to watch the race from the infield, with access to a trailer and other amenities. It was exciting and interesting, and I am grateful for having been included in these trips.

✄　◈　✄　◈　✄

Before I go much farther in this tale of places I have been I must tell about a group of friends who were sometimes called the "birthday girls." We lived in different parts of the state and had become acquainted by common interests in teaching and in Delta Kappa Gamma. The composition of the group varied somewhat at first, but settled down to seven in the early seventies. They were Ann Little Masemore, Mozelle Causey, sisters Mary Ella and Bertha Cooper, Rosalie Andrews, Phebe Emmons and myself.

Our designation as birthday girls came about because we got together for the birthday of each one, usually in the home of one of us. We also went together for a week at the beach, often in the Nags Head area, each summer. Then because some of us had family traditions which obligated us on Christmas Day, we celebrated Christmas together on New Year's Eve. We were an unusually congenial group. Nothing lasts forever, however, and in 1987 our plans for renting a cottage at the beach had to be canceled because of illness of three of the group. Two of those died within the next year, and two others within the next two years.

Cecil and I had always spent Christmas Day at home, visiting our families just before or after the day. I decided after he died that I never wanted to be at home on Christmas Day again. In 1970 I went to Richard and Clara's home and then on to Phebe's in Raleigh. Ever since then, except for 1971 and 1974, Phebe and I have planned a long or a short trip either including Christmas Day or soon after.

In 1972 the Delta Kappa Gamma Convention was held in Houston. Five of our group of seven drove to the meeting, and then went on to San Antonio. We left the car there with a friend, and took a flight to Mexico. We had several sightseeing tours in the area of Mexico City, including the Pyramids of the Sun and the Moon and the Shrine of Guadeloupe. We then took a plane to Guadalajara, where we had a delightful stay in an unusual sort of hotel called El Tapitio. There were several buildings on a hillside with lovely views of the Sierra Madre Mountains. A fancy little bus was available if we wanted to go down to the reservation area. Breakfast was brought to the room, and all meals were delicious. We took tours to Tlaquepaque, Lake Chapala, and the city of Guadalajara itself. One memorable sight in the city is the mural in the government building painted by Jose Orazco. When we returned to San Antonio our friend gave us a tour of that city, including, of course, the Alamo.

Part II
Some Serious Travel

During the spring of 1973 Mary Ella told me of a trip she and a friend were planning with a group led by a minister in her area. It was to include several countries in Europe. I was interested, but realized that the time conflicted with graduation day at Methodist College. With some trepidation I asked the dean at a committee meeting one day if I could be excused from the graduation exercise. He said it would be all right.

At that same meeting, Georgia Mullen, college librarian, was present, and as we were leaving the meeting she asked about the trip and asked whether or not I had a roommate. This conversation started a series of trips which have taken us to far-flung parts of the world. We have been together on thirty-two trips of at least one week in length, and twenty-eight of them have been outside the United States. We have visited forty-nine countries, some of them more than once, and thirteen Caribbean islands. While I'm going into statistics, I will also mention that from 1972 to the present, Phebe and I have taken twelve trips out of the country, and have had thousands of miles of driving and flying together within the United States. In addition, I have taken ten trips with another companion or as a single with a tour group.

I have kept a record of each year's travels beginning with 1973. To continue this story year by year would get too complicated, so it seems more logical to take it up by continents or other areas. So let's start with the United States and proceed to other parts of North America, then to Europe, Australia, Asia, Africa, South America, and Antarctica.

North America

✵ United States: 1973 to 1994

It might be well to concentrate first on North Carolina. My most frequent drive is to Lewisville to celebrate Christmas, birthdays or other special occasions, or just visit. The second most frequent trip is to Raleigh, where I go for meetings, concerts, or to visit Phebe. I believe that I have been in every one of the one hundred counties at some time or another for a meeting, to visit someone, or just pass through. I've seen the highway system develop from the time one drove through every little town to the present day when I can go to Lewisville without going through Winston-Salem.

My favorite place, besides home and Lewisville, is Chapel Hill; my favorite hotel is Grove Park Inn; my favorite drive is down the mountain (with someone else at the wheel) from Roaring Gap when you get a view of Pilot Mountain. There are many beautiful views and interesting historical places which I like to visit over and over, and the zoo is really special. I have also been to Shatley Springs, Hanging Dog, Elf, Why Not and the Nine Mile.

✵　　✵　　✵　　✵　　✵

Moving on to other states, I have attended Delta Kappa Gamma conventions in several of the major cities. From the first one in New York in 1968 to the one in Nashville in 1994, the most memorable one was in Seattle in 1974. This was true chiefly because Phebe was serving

as international president, and the North
Carolina delegation was responsible for
arranging a special breakfast and a reception. Following the convention, some of us
went to Mt. Ranier, then for a three-day trip
to Vancouver and Victoria. Of course the most
beautiful sight was the Butchart Gardens.

Another favorite convention was in Las
Vegas in the MGM Grand Hotel. I didn't
spend much money on the gambling
machines, but a group of us did see the
spectacular show, *The Titanic*. I also took a day off for a
bus tour to Hoover Dam. We were shown a film picturing
the entire story of the construction of the dam and the
power plant, and then were given a guided tour of the
facilities. It was almost unbelievable–I think of it as an
engineering miracle.

Space Needle–Seattle

In general, the Delta Kappa Gamma Southeast
Regional Conferences have been pleasurable and educational, but in the case of two of them, we remember something less enjoyable. The hostess state always plans a special "state night" and when we met in Memphis in 1981,
it was a Mississippi steamboat ride. The group I was with
must have been the last to get on the boat, because the
only seats left were just in front of a loud speaker.
Deafeningly loud music blared out the entire time. I think
we probably did not hear the announcement that dinner
was being served, and by the time we got to the buffet all
the food was gone.

The 1990 regional meeting was in Kissimmee, Florida.
Everything was all right except for the drive home. We
started out on a hot Sunday morning hoping to get home
the same day. Phebe was driving my car, and as the day

41

got hotter and hotter, we had the thermostat on low and the fan on high. Along in the afternoon a few miles south of Florence, South Carolina, the air-conditioning system started blasting out hot air. We were near a rest stop, so we drove in. Everything was closed for Sunday. At several intervals we tried to start the car, with no results. I decided it was time to use my AAA card. I used all the change we had and got only a lot of runarounds from whomever answered the phone.

Finally, two fellows, one of whom said he was the rest area security person (although he did not have a key to the rest room), came up and assessed our situation. His car was at the southbound side of the rest area but for a price his friend would take him to get it. He had jumper cables, and we got started. We drove with the windows open to Florence and got a room for the night. We had an uneasy rest, but the next morning the car started. We got home without risking the air-conditioning, but my service place found no trouble with the battery or the air-conditioning. I carry jumper cables now, and have canceled my AAA membership.

I mentioned that on the way from Kissimmee, Phebe was driving my car. This was usually the arrangement when we went on automobile trips together. I have developed some confidence in my driving through the years, and don't mind driving alone, but I would rather furnish the car and let someone else drive when it is possible. I had my confidence a little bit shaken when Phebe and I were coming through Tennessee one day on the way home from some meeting. She was not feeling well, and several times I offered to drive, but she always said she was all right. Finally I realized that she was quite uncomfortable, so I explained to her that it was my car and that

I could drive it. Her reply was, "I'm not that sick." I believe that the only time I ever drove when she was in the car was on a trip to the beach when she had a broken arm.

During and after Phebe's two years as president she was often called to speak at various Delta Kappa Gamma meetings. Early in March of 1975 she was asked to go to Juneau for the Alaska state convention. Since my spring break at Methodist College was at the same time I accepted her invitation to go along. However, I said that if I were going to Alaska I wanted to see more of the state than Juneau. Indeed, I wanted to go north of the Arctic Circle. Phebe's travel agent reported that planes would go no farther north than Kotzebue (just north of the circle) at that time of the year.

So, we planned a one-day flight from Fairbanks to Kotzebue, then back by Nome and Fairbanks to Anchorage. The flight to Kotzebue had one number, and the return flight had another, so we had to claim our bags and recheck them. This took most of our time in Kotzebue, but it didn't really matter, because there was about three feet of snow. Nome was also snowed in but there was a little more to see there. During the flight we had beautiful views of snow-covered mountains. We arrived in Anchorage late in the afternoon, and truly enjoyed our dinner of reindeer sausage and other delicacies. All day long during the flights our obliging attendant had served nothing but Oreo cookies and Coca-Cola. She brought these around very often, but never varied, and we could not locate a snack bar at either airport. To this day I avoid Oreos. The convention in Juneau was fine, and the Mendenhall Glacier is beautiful.

Another venture with Phebe was to the Northwest regional meeting at Bismarck. The treat which the North Dakota members had planned was a bus trip to Medora. This is a reconstructed pioneer village whose claim to fame is chiefly that Theodore Roosevelt had come there (at least once) on a hunting trip. We were entertained by presentation of an historical drama in an open-air theater. Another event of this conference was a trip to the Peace Garden on the Canadian border.

Several years later I went with Phebe to the California state convention in Oakland. We went a few days early so that we could spend some time in San Francisco.

Prior to each international convention, Delta Kappa Gamma sponsors a two-day conference called "A Seminar in Purposeful Living." Topics to be explored are selected with the hope that participants will get a new outlook on some aspect of life which is not a part of their usual activity and concentration. Phebe and I have attended six of these seminars, four of which were in the United States.

The seminar which preceded the 1994 convention in Nashville, Tennessee, was held in Asheville, North Carolina. Its emphasis was upon the creative arts, and we visited several craftsmen in their workshops scattered over the surrounding mountain areas. There were also sessions on mountain music and literature. From Asheville we drove on to Opryland Hotel in Nashville. The most memorable part of the trip, however, took place before we got to Asheville.

We were riding in my car, with Phebe driving along Interstate 40 between Morganton and Asheville, going sixty-five miles an hour, when POW! The left back tire blew out. Phebe pulled over very smoothly, and we looked at one another for a moment, stunned. Then I decided to get out and take a look, and about that time a young man was walking toward us on the shoulder of the road. As he came up he said, "If you have a spare, I'll change it for you."

So after removing our ponderous luggage from the trunk, he changed the tire. He had quite a difficult job getting some of the bolts loose, but he persisted. (Did I say that this was in the middle of a hot July day?) After he finished, I gave him a twenty dollar bill, and he said, "You don't have to do this."

"Yes, I do," I said, "and I think the Lord sent you to us."

He looked at the money and said, "I think He sent me to you." Then he asked if he could ride to the next interchange with us, adding that he thought we ought to stop anyway and see that the spare had plenty of air.

On the way he told us that he lives in Asheville and is a painter. He had been to Hickory to do some work there, but the person he was to see had left, and so he was having to walk back home. We told him we would be glad to have him go along with us. When we got off the interstate in Asheville we began to notice that the car was not running smoothly, and finally at a stoplight our young man told us he thought the spare tire had gone flat.

Would you believe that at that moment we were directly in front of a tire store? By the time we pulled in, the tire was beyond repair, so we told the fellows there that we would have to have two new tires. Out came all the luggage again, and a lot of computer-punching went on to get the proper tires. We told the man in charge about our previous misfortune and how lucky we had been to meet the young man who helped us. Then, being at their door when the other tire went flat was more evidence that

Somebody was looking out for us. I added that we would not have expected to find a tire shop open at six-thirty in the evening. He replied, "Well, our usual closing time is five-thirty, but today we had some extra work and had stayed open late."

Among the approximately dozen places in the United States in which Phebe and I have spent Christmas together, several stand out as especially memorable. One was Norristown, Pennsylvania, where we visited friends and were treated to a tour of Valley Forge. It was especially appropriate to try to picture Washington there in winter, because there was lots of snow. Christmas in Santa Fe is remembered for the ride through the city on Christmas Eve to see the thousands of luminarias, and the snow which greeted us on Christmas Day. In New York we, of course, saw the performance in the Music Hall, and several other shows. In Orlando we had a stay at the Hotel Grand Cypress and spent Christmas Day visiting the wonders of Epcot. A few years later we flew to Key West and rented a car so we could explore that part of the world. One attraction was the Hemingway House, with the many-toed cats. Another was the museum where treasures of the sunken Spanish ship, *Atocha*, were displayed. There were unbelievable amounts of silver and gold, jewelry and other items of interest. It is no wonder that the ship sank.

In 1984 I took advantage of Close-up Foundation's offer to have a session for older citizens. Although I had been in Washington several times I had never before had an opportunity to visit a session of the House (with Tip O'Neill presiding), to sit in on a committee meeting, or to have a guided tour of the Pentagon. The group sessions were thought-provoking and interesting. It was an enlightening week.

While on the subject of Washington, in 1990 Georgia and her daughter Kathy and I took part in a Christmas tour of Washington, Williamsburg, Charlottesville, Richmond and other nearby points of interest. As one of the special treats, I had never been inside the White House before, and this time was able to see it decorated for Christmas.

※   ✠   ※   ✠   ※

One of the other escorted tours Georgia and I have had in the United States was called "The Great West." It started in Denver, and from there we had a beautiful drive by bus through the Rocky Mountain National Park and on to Cheyenne and to Fort Laramie. Along the way there were several stops, and our guide furnished interesting background information. South Dakota was next, with Mount Rushmore as the main attraction. We were also shown the model and the beginning stages of a proposed mountain-sized statue of Crazy Horse. A ride through the game preserve in Custer National Park took us very close to bison, bears and other animals native to the region. Then, of course, we saw some of the area appropriately called the Badlands.

Back in Wyoming, we spent a couple of days in Yellowstone Park and visited the popular spots such as the geysers, waterfalls and the Devil's Tower. Old Faithful performed on schedule, but not with as much vigor as we had expected. We had beautiful views of the Grand Tetons on our way to Jackson Hole to see a rodeo.

The last stop before returning home was Salt Lake City. Parts of it had not changed much since I was there in 1936, but we got to see the home of Brigham Young and his many wives and children. An unexpected treat was to witness a rehearsal of the Mormon Tabernacle Choir, as well as the performance as it was being televised.

✄　✄　✄　✄　✄

My friend Ann Little was eighty-five years old in 1985. We had often heard her say that she had been in all the states of the United States except Alaska, and that she would not die happy unless she could go there. She was not physically able to go alone, so I decided to help make her wish come true and also visit some of Alaska that I had not seen. We planned a cruise up the inland passage, with a return flight from Anchorage. It was a beautiful trip, with stops at Ketchikan to see the totem poles, at Juneau for a salmon barbecue, and at Skagway to see the salmon research station. The weather was perfect for viewing the glaciers and ice-covered mountains, and we enjoyed it immensely. When Ann Little died at age ninety, I was gratified to realize that her wish to see all the states had been fulfilled.

✄　✄　✄　✄　✄

In 1987 Georgia and I took a trip to New England to see the fall foliage and then continue for a short distance into Canada. It started in Boston and continued into the White Mountains of New Hampshire. The mountains were beautiful with the brilliantly colored trees. On to Vermont, for more great views, and a visit to the Shelburne Museum of historic life in America. There were wonderful exhibits, of which my favorite was the display of quilts. We then crossed the border into Canada, first stopping at Montreal. On to Quebec, we had a brief visit to the Shrine of St. Anne de Beaupre. In Quebec we strolled around the narrow streets with views of the Citadel and the Plains of Abraham. We returned to the United States by way of Maine, and visited Acadia National Park and Bar Harbor.

❈   ❈   ❈   ❈   ❈

I have learned a little about New York from a few visits to the city. In addition to those already mentioned, Georgia and I took my niece Terri and her sister-in-law Ellen (ages in the early 20s) under our wings for their first visit there in 1982. We soon realized that we didn't have to guide them very much, because they quickly caught on to the layout of the city and of its transportation systems. We did the usual sightseeing things, and saw a performance of *Woman of the Year* starring Lauren Bacall. Georgia's son Tom, a West Point graduate who was serving in the army, met us and we rented a car and drove to West Point. With Tom as a guide, we had a most interesting tour.

More recently my trips to New York have been to visit a former Richlands student, Eldon Sanders, and his wife Anne. It has become an annual trip for the last six years, and Eldon and Anne have taken me to museums, church functions, concerts and plays and other interesting events. The outstanding happening of 1993 was a concert by Pavarotti in Central Park.

❈   ❈   ❈   ❈   ❈

As I recalled my 1936 visit to Yosemite, I thought now and then that I should go back and see if it was really as spectacular as I had pictured it. So, in 1994, when a pamphlet came announcing a trip which would include a brief stay in San Francisco, the wine country, Yosemite, Monterey, then a drive down the coast to Los Angeles and a visit to the San Diego Zoo, I decided to go. There was no problem in convincing Georgia to go, since she has a son and a daughter-in-law living near Monterey.

The day in San Francisco included a cruise in the bay, with a great view from under the Golden Gate Bridge and a circle around Alcatraz. An interesting account of the history of the area was available by use of earphones with a choice of translation into several different languages. Quite up-to-date and efficient!

We went through Napa Valley, but did our sightseeing of the wine country in Sonoma. There was a tour through the vineyards and wineries, with a stop for sampling now and then. Transportation was by rubber-tired wagons drawn by beautiful Belgian horses. One of the stops was at the Smothers Brothers winery. Our guide told us that those fellows had a hard time convincing people that they were in serious business, but that their wines have been given high ratings by professional tasters. It was also reported that on a television interview someone asked Dick Smothers how his wine business in Napa was going. He told them he was in Sonoma instead of Napa and added that "Sonoma makes wine, Napa makes automobile parts." T-shirts with this quip on them were available.

One of Sonoma's claims to fame is that it is the site of the world pillow-fighting championship competitions. In this game, competitors sit facing one another on poles set up across a big mud puddle. Each one pounds the other with a sturdy feather pillow until one falls into the mud. It will probably never replace soccer as the most popular game in the world, but we were assured that they have had participants from several other countries.

Yosemite Valley is still a magnificent place. We spent a night in the elegant Ahrahnee Hotel, with a spectacular view from our window. However, instead of driving around at will in a private car, as we had done when I was there before, we rode on a large open truck bed (with seats, of course) and stayed on prescribed roads which did not take us very near the places of interest. We had great views of El Capitan, but never had a good shot at Half-Dome. Much of our guide's message to us consisted of pointing out trails which we should come back later and walk. He also assured us that if we had "the time and the means" we would be able to go up to Glacier Point.

The greatest disappointment was that the waterfalls were bone-dry. It was almost funny to have the guide say we were approaching the beautiful Yosemite Falls, so many feet higher than Niagara, and then see a dry trench down the side of a high cliff. Our tour did not take us out of the park to the redwood forest, although later on our bus took us off of the coastal highway a short distance in it and we strolled around and marveled at the girth and height of those trees. I read also that the Tunnel Tree, through which we had driven in 1936, had blown over in 1968. So, I guess you can't "go back again." However, I still contend that Yosemite is worth the trip, but go in the spring or early summer when the waterfalls are still being fed by the melting snow in the mountains, stay long enough to walk on some of the trails and go up to Glacier Point for a breathtaking view of the valley.

Our only two-night stop was at Monterey, and Georgia spent the first night with her son Bill and his wife Norma. Our tour group spent the next morning at the Monterey Aquarium. This is an absolutely marvelous display of organisms which live in the water. There are exhibits of the many habitat types, with not only fish, but also shellfish, corals, sponges and the plants which live in that particular environment. One area which especially attracts children was the open "hands-on" pool, featuring bat-rays (big black fish shaped like our stingrays). The most impressive exhibition was a three-story-high kelp forest.

Georgia and Bill and Norma picked me up at noon, and we had a beautiful drive along the Big Sur. We had several leisurely stops for walking off of the highway to get closer looks at some of the special points of interest. The next day our bus took us over much of the same highway, but our stops were few and hasty. This is a magnificent, although very curving, drive along the cliffs which face the ocean.

On our way we had a visit to the W. R. Hurst Castle. While the evidences of wealth and extravagance were impressive, this place did not represent the kind of lifestyle we would wish to have.

Near Los Angeles in Malibu we visited the J. Paul Getty Museum. The building is said to be a replica of a seaside villa in Herculaneum which was buried by the eruption of Vesuvius in A.D. 79 and partially uncovered in 1709. The most extensive displays consisted of ancient Greek and Roman art objects, but there were also large areas devoted to paintings and sculptures by the old masters and by other artists up to the nineteenth century. What we had time to see was very impressive.

Our overnight accommodations in the Los Angeles area were at the Beverly Hills Hilton. Although it was billed as the "meeting place of the stars" and was very large and imposing-looking, we didn't see any stars. Actually, it seemed rather unfriendly to a busload of tourists.

In Los Angeles itself, we did not go into the areas of strife and poverty, but stayed in the downtown area. The streets were wide, with lots of trees and flowers, and everything was peaceful. I remembered that when I was there previously there were no buildings higher than three stories because of the threat of earthquakes. Although that threat still exists very tall buildings are possible because of the improved methods of construction.

We drove a long way on Hollywood Boulevard and stopped at the only place which looked familiar to me— Grauman's (now Mann's) Chinese Theater. This theater opened in 1927 with the first showing of the movie *King of Kings*. For a number of years the Academy Awards were presented there. It is probably most famous now for its plaza in which stars through the years have put their hand and footprints in wet cement. Although the area is many times larger than it was in 1936, I was glad to be able, with help from fellow tourists, to find the square in which in 1929 Tom Mix had put his prints along with the hoof prints of his horse Tony. Among others dating back to those days were Mary Pickford, Norma Swanson and Eddie Cantor.

On our stop at the mission of San Juan Capistrano we saw lots of pigeons, but no swallows. The mission is beautiful and we had a leisurely walk to enjoy it.

The San Diego Zoo tour was reminiscent of the one in Yosemite—riding on a bus, seeing very little, and getting suggestions about places to which we should walk and get a better view. We did happen to see a few animals, and it was interesting to get an idea of the layout and the immensity of the place. To see it properly would require several days.

✸ *Canada: 1980, 1986, 1990, 1994*

In 1980 the Delta Kappa Gamma Convention was held in Detroit, and the pre-convention seminar took place in Ottawa, Canada. Phebe and I flew to Ottawa, and had made reservations for a bus from there to Detroit. I feel sure that the sessions were inspirational and uplifting, but what I remember best was that a lady who was Minister of Health and Welfare came to greet us and invited us to a reception in the Parliament Building. It was a lovely reception, and following it we observed a part of a session of Parliament. It was especially interesting to have the speakers alternate from French to English. Another highlight of our stay in Ottawa was an opportunity to see a performance of *La Bohéme* at the beautiful opera house.

We had arranged our bus tickets so that there was a stop in Stratford, where we were to see a Shakespearean play. On an early Saturday morning we got to Toronto, where a bus change was required. Right away we found out that the early bus to Stratford did not run on Saturday. We could get another bus in about four hours, if there were seats available. Meanwhile we were warned that we must not take a chance of missing this opportunity by leaving the station. So, we spent the morning in the unattractive and uncomfortable Toronto bus station, but did get to Stratford with minutes to spare for the performance of *Twelfth Night.*

The next part of this episode is that on the way to Detroit we were stopped at the border and all of our luggage was taken off and thoroughly searched. Then we personally were searched and questioned. We never found out what they were looking for or if they found it, but this was an unexpected thing at a border where the question, "Are you a United States citizen?" is usually sufficient.

❋　❋　❋　❋　❋

Georgia and her daughter Pat and I took a tour to Canada for the 1986 Expo in Vancouver, with continuation into the Canadian Rockies. The Expo was interesting and well-organized, but the rest of the trip overshadowed it. I had heard rave reviews of the scenery in that part of the world, and it came up to expectations. The many lakes reflecting the surrounding snow-covered mountains were unbelievably beautiful. My favorite was Emerald Lake, which is really green. Mt. Robson, the highest of the peaks, was unusually impressive as it sort of stood alone, with no lesser peaks crowding around it. We were actually permitted to walk for a short distance on a glacier, the Athabaska.

Along the way one day we stopped at a ridge which was described as the Continental Divide of Canada. Water flowing on one side of the ridge separated, and some of it went west to the Pacific and the other went by a circuitous route to the Atlantic, according to what we were told. There actually were two well-defined shallow ditches going in opposite directions downhill. Our guide admitted that a little man-made manipulation had probably been used but I was glad to see this so definite, because in crossing Colorado, where a sign says, "Continental Divide," the area looked completely flat to me.

At our last stop in Canada, Lake Okanaga, I saw a placard describing a nature trail not far from the hotel. A visit to this trail was not on our agenda, but I got up early and did a private tour of part of the trail. Signs along the way told of the geology and gave the names of some of the plants. There were beautiful scenic views and this was the high point of that day for me.

❊   ❊   ❊   ❊   ❊

Four years later, Georgia and I decided to go back to Canada. We planned two two-week trips, with less than a week between. The first one was to Nova Scotia, Prince Edward Island and New Brunswick. We joined the tour group in Boston, and went by bus up the coast. This time we made a side trip to Kennebunkport to get a look at the home of its most famous residents. From there we went to Portland and took an overnight ferry across the Bay of Fundy to Yarmouth.

The next morning we boarded a bus and drove up the coast of Nova Scotia. The first attraction was the land of the Acadians, whose story was reviewed by our guide. We were traveling with RFD tours, and although they had expanded their clientele, they were originally formed to serve rural people. True to their origin, they had us stop to visit a dairy farm. We watched a flag-raising ceremony at the Citadel, near Halifax, then went on to Peggy's Cove. From there we saw the Alexander Graham Bell museum, and talked over one of his early telephones.

Prince Edward Island claims to be the chief potato raising area in the world, so we visited a potato packing house. I bought a recipe book which gives instructions for making an entire six-course dinner with foods containing potatoes. I tried this and it was a pretty good dinner. Another stop was at the house which inspired the writing of *Anne of Green Gables*.

The most appealing lure to this part of Canada was the hope of seeing the tide come in or out of the Bay of Fundy. This was promised to us in New Brunswick, but although we saw many areas at low tide we never saw the action. The nearest we came to it was at the Reversing Falls. In the evening the tide was coming in and in the morning it was going out (or vice versa) but the difference in level of the water didn't create much of a fall. On our way back to the United States we stopped at the home of Franklin D. Roosevelt at Campobello, now a museum.

❊   ❊   ❊   ❊   ❊

Our second trip to Canada started at Fargo, North Dakota, and then to Manitoba. This was also by RFD, and our farm visit was to a community of Hutterites, near Winnipeg. This group descended from a religious and social sect exiled from Germany and then from the United States because of their beliefs. This colony seems to be thriving in Canada. They are self-sufficient in many ways, and use up-to-date farming machinery and methods. They obviously have extensive markets for their products, as they had, among other things, three thousand geese and hordes of clean white pigs. We had lunch with these people, and complied with their custom of having men and women sit at opposite sides of the room.

This trip was called "Canadian Wildlife," and we went on several excursions to a place called Riding Mountain National Park. The scenery was beautiful, and we saw a good many birds, but not many animals. The chief drawing card for this trip had been the promise that we would see many, many polar bears at Churchill, on Hudson Bay. Our first day there was very cold, and we spent it visiting an old British fort. The next day a heat wave set in, and from then on we had temperatures in the nineties. Of course no self-respecting polar bear would be caught out in this kind of weather. We rode across the tundra in an interesting vehicle called a "tundra buggy," and saw lots of birds, but no bears. We had better luck at spotting beluga whales on a boat trip.

The thing which made the trip truly memorable was that we went away from the bright lights one night and saw a magnificent display of aurora borealis. It was mostly white, with changing waves of pink, blue, and green, filling the entire sky and moving like ethereal curtains. We watched for a long time. It was an awesome sight.

❈ ❈ ❈ ❈ ❈

Gwen Simmons and I had a trip to the Canadian Rockies in May of 1994. In 1980 I had seen much of the territory we visited, but this trip began in Calgary, which was new to me. Although it is known chiefly for its "stampede" (a sort of glorified rodeo), it is a busy center of the Canadian oil industry. Most of the buildings are relatively new, there are lots of trees and flowers, and not a sign of litter or graffiti. We were told that there is practically no unemployment, no slums, and a very low crime rate.

On our bus rides through the Rockies we were fortunate in having beautiful weather. Views of the mountains, lakes, and glaciers were spectacular. Emerald Lake and the sapphire-blue Lake Peyto were just as beautiful as I remembered them, and the tall spruce trees were as magnificent as ever. The Continental Divide had been given greater emphasis. Lots of rocks had been added to outline it, and there were steps and railings leading to it.

A real innovation was a snowmobile ride up to a high level on the Athabasca Glacier. There was a steep climb on a road which had been cut out of the ice and snow. We understand that the road has to be relocated every seven years so that the glacier can rebuild itself. We got out and walked on several hundred feet of ice at the highest ice fall before the head fall. It was very cold and windy, and, appropriately, it began to snow.

We saw lots of animals in their native habitats—several near enough to get good pictures of them. There were longhorn sheep, mountain goats, elk, one lone black bear, a coyote, many ground squirrels, and, for the first time in my life, I saw a pika.

Rocky Mountain Sheep

Mexico: 1974, 1991, 1992

For Christmas of 1974, Mozelle and I decided to take a Delta Kappa Gamma trip to Mexico. We met at Greensboro Airport and waited for several hours for a plane which was held up by snow in Boston. Finally we got to Atlanta, already certain that our plane to San Antonio had long since departed. We had our names listed for a much later flight, but were dismayed to see that the official in charge had a stack of requests about six inches thick. He peeled off a few at a time as he learned of an available seat. Finally he came to us, but had only one seat. We must have been living right, however, because someone realized that a couple whose names had not been called had got on the plane. The attendant at the hotel in San Antonio was somewhat startled when we checked in at 4:30 A.M. and asked for a 6:00 A.M. wake-up call.

A special event for Mozelle and me in Mexico City was an invitation to visit the president of the Delta Kappa Gamma chapter there. She sent her chauffeur for us, and when our visit was ending she asked him to drive us around the city to see the beautifully lighted Christmas decorations. We spent Christmas Day at the archeological museum, with its wonderful displays of life in Mexico from the earliest known era to the present.

❉   ❉   ❉   ❉   ❉

Copper Canyon, located in the western Sierra Madre Mountains, is said to be four times as extensive as our Grand Canyon. There are no highways into the area, and the only way to get there is by train (or on mule-back or on foot). In December of 1991, Georgia and I met with a group in Tucson and went from there to board the Sierra Madre Express in Nogales. The route followed the river for a while and then rose to a height of eighty-five hundred feet to the rim of the canyon. During the ascent the train went through eighty-five tunnels and crossed thirty-six bridges. Views along the way were spectacular. We spent the first night on the train, and another one on the return to Nogales.

Meanwhile our cars were disconnected for a day at each of five small settlements along the way. We spent the nights in the local hotels, which did not offer first-class accommodations but were clean and adequate. During daylight hours at these stops we took bus or walking trips to get better views of the canyon. At each spot the views were strikingly different, but always amazing. The train picked us up on its way the next day and took us to the next stop.

Much of the area is inhabited by the Tarahumara Indians. Groups of women and children seemed to know of every stop we were to make, and were there ahead of us to show us their woven baskets. They knew little Spanish, but could tell us the price of the baskets in pesos. Sometimes they also had dolls and jewelry made of beads. We learned that the Indians lived in caves along the walls of the canyon. Their chief food seemed to be corn, which they grow in small patches where soil is available. Evidently the women do all the work and the men engage themselves in drinking one of the products of the corn.

❊　❈　❊　❈　❊

During the Copper Canyon trip we learned of another trip, "Colonial Mexico," which RFD had planned for the coming year, and decided to take it. On the first day in Mexico City we were treated to an outstanding performance by the Ballet Folklorica in the impressive Fine Arts Theater. We then visited several sites in the city, including the square in which the Aztec capital stood until it was destroyed by Cortez.

The next day we went to the Shrine of Guadeloupe. When I was there in 1974 the original church had sunk so much that it had to be replaced. The new and larger one was very beautiful inside and out, and was of the same architecture as the first one. By 1992 the second one had also been replaced because it had started to sink, but both of the old ones were still standing, looking very forlorn. An enormous new basilica, of "modern" style of architecture, had been built as a shrine to house the shepherd's blanket with the picture of the Madonna. We were told that a new type of underground foundation assured that the new one would not sink.

The Teotichuacan temples and pyramids were much as I remembered them, except that some of the walkways and walls were being restored. One place of interest was the city of Dolores Hidalgo, where Father Hidalgo and Miguel Allendo plotted the first attempt to overcome the Spaniards in Mexico. In Guanajuato we saw the setting of that battle. On the way to Guadalajara for our flight home we visited a silver mine, saw a lively Indian dance festival, and had a boat ride across Lake Patzcuaro to the island home of the Tarascan Indians.

Teotichuacan Pyramid of the Sun—Mexico

The Caribbean: 1974, 1975, 1985, 1986

During Methodist College's spring break in 1974 some of the members of the Science Club planned a trip to Jamaica. The parents of one of the students were connected with a family which owned a villa on Montego Bay, and they were able to rent it for us and to go along. Seven students and I joined them. The villa came with a cook, a maid and a gardener who looked after the grounds and the pool.

We rented two cars soon after we arrived, and it was amazing how quickly two of the students adapted to driving on the "wrong" side of the road. We rode around a good bit and asked a lot of questions of the residents, who were very cordial and helpful.

We visited a private bird sanctuary where many birds came to feed, and were able to entice hummingbirds to perch on our finger and sip sugar water from a bottle. We took a train trip into the Blue Mountains, attended a flower show, investigated a rum distillery, and had a taste of night life where one of our group made a good showing in the limbo contest. Our cook took us with her to the market, where she bought goat meat (for curried goat), codfish and akee (the national dish), soursop (for a sweetened beverage), and other foods which were strange to us but delicious.

The greatest attraction was in the water. We could wade in the shallow edge of the bay and see starfish, sea urchins and anemones, peacock worms, crabs, and other sea life. Some of the fellows had brought their scuba equipment, and went out farther to the coral reefs. All of us had a ride in a glass-bottom boat.

One of the girls, Gail Worth, had a saltwater aquarium at home and was eager to take home some peacock worms. Since we could not find anything in customs regulations which prohibited saltwater animals, Gail put some sea water and a peacock worm in each of four half-gallon ice cream cartons. She carried these in a large tote bag and got them safely to the airport in Atlanta. Of course she had to explain what she was carrying to the customs agent, and he was in a quandary. After making several phone calls he sent her to a private room with another agent. After what seemed like a long time, he came out and a lady went in. Another long wait, and Gail came out, carrying her worms. She reported that everything in her luggage had been examined minutely, and then she had been asked to completely undress (yes, this was after the woman took over). We decided that these people felt sure that anyone who would admit to carrying peacock worms must be hiding something illegal. Incidentally, a peacock worm is a marine animal which is pretty homely-looking until it occasionally spreads out its gills, which look something like a colorless peacock tail.

✂ ✠ ✂ ✠ ✂

During the week after Christmas in 1975, Phebe, Mozelle, and another friend and I celebrated the season with a Caribbean cruise. We went to Ocho Rios, a place I had missed in Jamaica on the other trip, and also visited a house which a local guide told us had been a voodoo center. We had another stop at Grand Cayman, which I don't remember much about except that it had beautiful beaches and we had turtle burgers to eat. Our third excursion was to visit the Mayan ruins on the Mexican island of Cozumel. The ruins were impressive, and our guide was well-informed. His English was a little shaky, however. At one point when we were headed toward a particular shrine there was a large crowd of people there, and our guide said, "We come back later—is too much pipples there now."

✂ ✠ ✂ ✠ ✂

My next venture into the Caribbean was through the Panama Canal in 1985. Actually, Georgia and I met our ship in Acapulco, and had a little sightseeing there. We later had a day off the ship for a tour of Costa Rica. We were impressed by the beauty of the country and by what we were told of its stability as compared to some of the other Central American countries.

Transit through the canal was an amazing experience. We had lectures reviewing the building of the canal, and information all along the way as to what was happening. It was so well done that even a person who knew practically nothing about engineering got some understanding of the procedure. On the Caribbean side we had stops at Cartajena and at Aruba and Curacao, then flew to Miami.

✶   ✤   ✶   ✤   ✶

I had often said that I wanted to live long enough to see Halley's Comet when it appeared in 1986, so as that time approached I began to study any travel literature which promised a good location for viewing it. Sun Line offered a cruise in the Caribbean, the chief object of which was to see the comet, and Georgia and I selected that trip. We were well prepared since we read all of the information we received. On the ship there were several scientists who lectured on a number of subjects relating to space in general and to comets in particular.

The plan was to maneuver the ship so that it would be away from lighted islands when the comet was at its best point for viewing. That turned out to be at about 4:30 A.M. We were asked to put a note on our door if we wished to be called. When our scouts found that the sky was clear, we stumbled up on deck in the dark. The comet was actually visible to the naked eye, but a pair of strong binoculars helped.

We had a side trip for some distance up the Orinoco River as far as the ship could go, and then had a flight farther inland. That was planned in order to provide an open dark space for those who had cameras which would turn as the earth rotated and take timed exposures. Unfortunately, it was cloudy that night. The next day we flew to a resort called Canaima, and on the way had a quick look at the highest waterfall in the world, Angel Falls. From the distance at which we saw it, it was not very impressive.

During the daylight hours each day we had a stop at one or two islands. They all began to look pretty much alike, but I remember Mt. Pelee on Martinique, the boiling sulfur springs on St. Lucia, banana trees everywhere, and a sign on a tree in Bequia which read: "No one is allowed on this breadfruit tree without the oder [*sic*] of Miss Pollard."

Europe

🔯 *First Alexander Tour, 1973*

My first trip to Europe except for the Russian tour in 1971 was arranged by David Alexander, a minister in Kinston. Our Swiss Air flight took us to Zurich, where we were met by Willi, a young German man who was a marvelous bus driver and a knowledgeable tour director and guide. We went first to Lucerne, where we took a city tour. It started with a walk across the Lake Lucerne Bridge, with its beautiful paintings overhead. We then rode around the lake, stopping at the statue of the lion of Lucerne and other points of interest. We had free time to wander on our own and to shop. Then the highlight of our stay was a cable car ride to the top of snow-covered Mt. Pilatus.

From Lucerne we rode through beautiful mountain areas, passing by villages with their steep pointed-roof houses and abundant flowers. At the St. Gotthard Pass, our bus was driven on a special train car and we rode piggyback through the tunnel into Italy. The first view of Italy was of fairly flat land, with hay fields and orchards and vineyards. Soon we came to the Appennine Mountains, which were not so high and rough as the Alps.

Our next destination was Florence, where the magnificent Cathedral Santa Marie de Flore is the outstanding sight. Work on this building began in 1296 and was not completed until 1471. The outer construction and the interior artwork are so marvelous as to defy description. One of the chapels was designed by Michelangelo and contained seven of his sculptures. The nearby baptistery is justly famous for the twelve embossed and gilded representations of Bible stories on its door.

Of course the most famous work of Michelangelo in Florence is the statue of *David*, at which we marveled in the Academy of Fine Arts. We walked across the Arno River on the Ponte Vecchio and enjoyed the beautiful art works and jewelry in the shops.

On the way to Rome we stopped at Assisi, which looks like a walled fortress on the side of a mountain. We spent most of our time in the Basilica of St. Francis. There were actually three churches, one built on top of the other. The lower level contained the crypt where St. Francis was buried. There were mementos of the saint, and paintings of his marriage to poverty and his preaching to the birds.

Most of the places of interest we saw in Rome were familiar to us from pictures, but it was very gratifying to see and walk among them and in them. There were the Colosseum, Victor Emmanuel Memorial, Pantheon, Spanish Steps, Trevi Fountain (we threw in some coins),

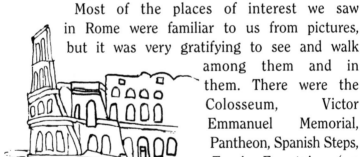

The Colosseum—Rome

and many other statues and monuments. Of course, the highlight was to go to the Vatican and see the Pieta and the ceiling of the Sistine Chapel, among other beautiful and significant things in St. Peter's Cathedral. Our hotel was across the street from the Church of St. Peter in Chains, and we walked across to see Michelangelo's *Moses*. Willi was sufficiently familiar with the city to drive us to all sorts of interesting views and we had some free time to go around on our own.

Venice was our next stop. This is the city of canals, and the only transportation is by walking and by boats. Georgia and I decided to go for a walk after dinner, and we got lost. Fortunately, we knew the name of our hotel and a kind lady led us to it. We learned later that there are no street names, and addresses are by districts. According to our local guide, "Only God and the mailman know where anyone lives." The next morning, Sunday, we took a boat to St. Mark's Square so that Georgia could attend mass at the cathedral. Later in the day we had a group tour and spent some time walking around the square, watching the glockenspiel and the multitudes of pigeons. We also had a tour of the magnificent Doges Palace, the Bridge of Sighs, and other points of interest.

On the way to Innsbruck the scenery changed from level farmland to mountains which became steeper by degrees. We were in the Dolomitic Alps, which were different from the Swiss Alps in that they had scattered patches of crumbly, rocky debris. By the time we reached Brenner Pass, the mountains looked more like those of Switzerland.

It was pretty much downhill to Innsbruck, nestled between mountains on the Inn River. We had rooms in the Maria Theresa Hotel, on Maria Theresa Street, and were near the imposing statue of Maria Theresa. One of the outstanding features of our city tour was the fantastically lovely Hofkirche, a Baroque style church dating back to the sixteenth century. Our tour also took us to the site of the 1964 Winter Olympics. The stadium, the ski jump and the Olympic symbol were still there. From this area there was a magnificent view of the city below, with snow-covered mountains behind it.

The road on the way to Munich followed the Inn River for a while and there were grazing cattle along the way. In Munich we saw another Olympic site, this time that of the summer games in 1972. The area was built upon an accumulation of rubbish from structures which had been destroyed during World War II. The apartment building in which several Jewish athletes had been killed was pointed out to us. On a happier note, we visited the Nyphenburg Palace with its exquisite furnishings and lovely gardens. Our favorite spot was the Marienplatz, a square where people gathered for refreshment and to watch the unique glockenspiel.

We finished our trip with a short tour of Worms, where the main points of interest were the statue of Martin Luther and his Church of the Reformation.

Another Christmas in Russia, 1973

Anyone who thinks it was ridiculous to go to Russia in the winter will surely think it is idiotic to go a second time. However, there were inducements. This trip was also sponsored by Delta Kappa Gamma, jointly with the University of Virginia. Two members of each group were given free trips. Phebe, as president, and a member of the headquarters staff were to head the Delta Kappa Gamma contingent. The other member decided at practically the last minute not to go, and I was asked to go as Phebe's traveling companion. Not only was it a free trip, but it included a first-class airline ticket.

The agenda was much like the one in 1971, but there were some aspects in which things were different. The atmosphere was less tense, and we were allowed more freedom to go places on our own. We had lectures by officials who told us that the people had more money than they could spend, because of the shortage of goods available from other countries, especially the United States. We saw evidence of this when we went to the big Gum department store and merged with crowds of people vying for the high-priced articles on sparsely filled shelves.

One big difference was that there were lots of automobiles, which had been produced in Russia. They were all the same model, black, and not very attractive, but

Kremlin Tower—Moscow

they could cause a traffic jam. The food was more varied than before—more meats and greater variety of vegetables. There was even ice cream and an occasional orange. Visits to schools and museums, and the sightseeing, were sufficiently different to be interesting, and a great many more trips would be required to see everything in the Hermitage and the Kremlin museums.

Russians ignored Christmas but for New Year's Day they decorated trees and gave gifts. We attended a festival of Father Frost and the Snow Maiden. Father Frost has a white beard, is dressed in blue with white fur on his cap and his long coat, and gives New Year's gifts. The Snow Maiden, his helper, is predictably dressed in white. Actually we didn't stay long at the festival, because it was held outdoors and it was very cold.

✠ *England-Scotland Study Tour, 1975*

During the summer of 1975 a group of Methodist Church officials in Florida planned a study tour for teachers in the southeastern states. There were about seventy-five participants, twenty-seven of which were from North Carolina and included Phebe and her Aunt Jinx Wood, Mozelle and myself. The academic study was for two weeks at the University of Edinburgh; this was followed by a week-long tour of England.

Our first night was spent in London, and a group of us, some of whom were familiar with the city, decided to take a taxi to Westminster Square. Everything appeared at first to be closed, but there were lights on in the Parliament Building and we saw people going in and out. We ventured over, and learned that we were permitted to go up into the Stranger's Gallery and observe the proceedings in the House of Commons. It looked just as it is pictured, with the presiding officers in their robes and white wigs. At one point the vote on a proposal was close, and the ruling was challenged. To settle this, all of the members walked out, and returned immediately. The count was taken as those who favored the proposal in question came in one door and those opposing it in another.

We went on to Edinburgh the next day by bus, passing fields with hedgerows separating them, and houses with chimney pots on their roofs. There was a brief stop at York to see the remains of an old Roman wall.

In Edinburgh we were given rooms in the Pollock Halls of Residence and a schedule of the lectures on "Comparative Education in the British Isles." Lectures were held in mornings and evenings, and a series of tours was available in the afternoons. We toured Edinburgh, including the castle which towers above the city and has a history dating back to the 900s. Many reminders of other times were on display, including the crown jewels of Scotland. Another famous building was St. Giles Cathedral, where John Knox founded the Presbyterian Church.

Other tours took us to the home of Sir Walter Scott, the Trossack Mountains, Stirling Castle, and the Linlithgow woolen mills and shops. Jinx and I spent one entire day in the Botanical Gardens, and did not see nearly all of it. One long lane was lined with large maple trees, and a sign informed us that every kind of maple in the world was represented there. I believed it when I spotted a Carolina Maple. The most unbelievable plant was a water lily, native to the Amazon, with leaves about six feet in diameter.

For the first weekend the entire group took a trip to Inverness. Along the way we saw that dry walls (rock walls which had no cement) divided the fields. There were lots of sheep, and when we stopped to eat our bag lunch some of them came to greet us and share our food. They loved our cheese sandwiches, which they would eat out of our hands. At the Spittal of Glenshee—honestly that is the name of the place—we stopped to watch a *tomintoul* (bagpipe band) on the village square.

In Inverness we saw the lake, but not the Loch Ness Monster. On the way back to Edinburgh we passed by Loch Lomond, and heard the legend relating to its bonnie banks.

Classes were canceled one day so that we could go to town to see Queen Elizabeth greet the newly-crowned King of Sweden, King Carl. The royal coaches and white horses had been brought from London, and we had gone early enough to get a good spot from which to see the parade with the Queen, the new King and other members of the royal families and staffs. It was quite sensational. Several days later we saw the notables again as a formal good-bye, with a military retreat, was said to King Carl at Holyrood Palace.

On our second weekend in Edinburgh the group tour was to Glasglow. Phebe, of Scotch descent and an alumna of Flora MacDonald College, wanted to go to the Isle of Skye, with which Flora was associated. Much maneuvering was necessary, but we were able to get a sleeper train to Inverness, a local train to Kyle of Localsh where a ferry connects with Skye, and overnight accommodations at Inverness for Saturday night.

At Localsh we learned that the passenger bus on the ferry did not run on Saturday. Having come this far, however, we decided to hire

Statue of Flora MacDonald
Inverness, Scotland

a car and driver to take us to Skye. He was very obliging, gave us much information about the island, and took us to the castle in which Flora MacDonald had lived for a while. When we got back to the mainland we had to pay him all the Scotch money we had, but we felt that it had been worth it. The only problem was that U.S. money and travelers checks were not acceptable in Localsh. We had our return tickets, and felt sure that our travelers checks would be good in Inverness. On the way, Phebe found a ten-pound note that she didn't know she had, but it didn't help much because when we got to Inverness all of the restaurants, including the "Ness Cafe," were closed. The kind receptionist at the hotel said we could have a sandwich in our room and that Leslie, the bellhop or whatever, would "pop up with it."

Back in England we stopped at Durham Cathedral, built in the eleventh century with additions throughout the years. It had a rose window "as long as a tennis court." On the floor near the back of the nave there was an arrow which originally marked the point beyond which no females were allowed to go.

We next visited the cathedral and a museum in York, and had an overnight stop at Leeds. On to Coventry we saw the modern cathedral which replaced the one destroyed in World War II. It was built by contributions from many nations, and tiles on the floor pictured each of the continents. The "world's largest tapestry" made by a group of women in France hung behind the altar. Near the entry there was a cross made from some burned timbers of the original building. While in Coventry we also sought out the statue of Lady Godiva on her horse. She really did have long hair!

Along the way the British version of some signs attracted our attention: "Exit" is "Way out," "Pass with care" is "Think before overtaking," and now and then there was a "Diversion ahead." Our driver pointed out to us several "ricing" (racing) courses.

A visit to Stratford-on-Avon gave us an opportunity to see several places of interest related to Shakespeare, including his birthplace, his tomb, and Anne Hathaway's cottage.

In London we had sufficient time in the British Museum to see the Rosetta Stone, the Magna Charta, the Elgin Marbles, and a quick look at the African exhibit. Next there was a tour of Westminster Abbey, where we saw the nave and a number of chapels; the tombs of several rulers including Elizabeth I and Mary, Queen of Scots; the coronation chair with the Stone of Scone under it; and the marble plates on the floor marking the burial places of such famous people as Charles Dickens, Handel, Isaac Newton and Charles Darwin. Our tour also took us to the Tower of London and a look at the crown jewels.

✠ Second Alexander Tour, 1976

Our trip to Denmark and Norway in 1976 began with some excitement in the foreign flight lounge of Kennedy Airport. We learned that Queen Margreth of Denmark, who had been visiting in the United States, was to be on the plane with us. We had a good look at her as she went shopping in the airport. Of course she and her entourage occupied the first-class section, but we saw her again as she greeted her family in Copenhagen.

We had some free time to walk around and shop, then the group had dinner at a lovely restaurant in Tivoli Gardens. A sightseeing tour of the city next day took us to the Little Mermaid statue, Parliament House, the Royal Palace, several churches, and so on. We heard a good bit about the history and the legends of Denmark.

The Little Mermaid

One legend concerned the magnificent Gefion Fountain, showing a woman plowing with three oxen. The story was that a goddess had been sent from heaven to be the wife of the Scandinavian king, and had borne three sons. When the time came for her to go back to heaven she asked to be able to bring some of the land back with her. She was told that she could bring all she could plow in one night. So she changed her sons to oxen and they plowed all night. However, the plot of land was too large and too heavy and she dropped it in her attempt to carry it to heaven. The dropped land became the island of Zeeland, and it just happens to have the shape of the largest lake in Sweden.

From Copenhagen we had a tour which took us across Zeeland Island, then by ferry to Funen, and another ferry to the Jutland peninsula. Across Zeeland we saw lots of hayfields and dairy cattle. Unbelievably, some of the cows were wearing a special kind of brassiere. This was probably partly for their comfort, but it also kept their enormous udders from dragging on the ground.

We had several stops along the way at castles, churches and other notable buildings. At one of the castles we were told that the king in residence at one time liked to impress visitors by having swans in the moat. Unfortunately, the water was so polluted that the swans soon died, so he had to order new ones every time he had guests. In one of the residences the beds were very short; this was all right because the people slept sitting up to keep from mussing their hair-dos. A room in one of the castles was completely wallpapered with tooled hide of wild boars.

We were told that the hotel in which we stayed in Randers was the site of the first public appearance of Victor Borge. I played a few chords on the piano which he is said to have used. Also in Randers we saw a stork's nest on a chimney, and the heads of two young ones could be spotted. We waited a long time for one of the parents to come and feed them, and joked about waiting for the stork.

In Funen the chief place of interest was the home of Hans Christian Andersen in Odense. The house had been made into a museum of articles relating to him.

For some reason, the people of Aarhaus have been the subject of jokes reminiscent of some we have used about Poles. We heard several of these and a short example is: "Why do the men of Aarhaus put on their best clothes during a lightning storm?" Answer, "Because they think their picture is being taken."

One of the men in our group realized that he had left his pajamas in the hotel in Copenhagen, and the guide called back and asked that they be forwarded to him. He received a message which he read to us: "If your pajamas exist, they will meet you at the dock." They did meet him, so we were able to take the ferry across the Kattegat to Norway.

After a brief stay in Oslo we took a flight to a small northern town named Alta. From there we had a bus ride to Honingsvaag, the most northern town in Europe. On the way we saw many reindeer, sometimes with their Lapp herdsmen. Our chief object was to see the sun at midnight, and as we were watching television in the hotel lounge one of the women in our group came by and said, "I'm going to bed, but wake me when the midnight sun comes up." We didn't have much success

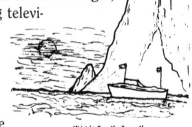

Midnight Sun—Nordkapp, Norway
June 20, 1976

explaining to her that it didn't just come up at midnight, but we stayed up and so did the sun. The next day we went to Nordkapp, the most northern point of land in Europe, and got a marvelous view of the sun at midnight reflected in the Arctic Ocean. Our guide told us that we must throw a stone into the ocean as a means of protection against trolls.

Back in Oslo we saw Frogner Park, the location of the outdoor museum of Vigeland sculptures. There are magnificent and enormous figures representing many aspects of life. One sixty-foot-high column showed figures of two hundred people of various ages, trying to reach the top. Some are succeeding, some helping others up, some pushing others down, and some are slipping down or falling. One statue which was brought to our attention showed a father, grandfather and son. Since David's father and teenage son were with us, it seemed appropriate to have them pose by the statue. Of course the figures were nude, so the grandson asked, "Do we have to take off our clothes?"

Our agenda also included a Viking museum and the Kon Tiki Museum, which displayed articles relating to Thor Heyerdahl.

Following our stay in Oslo we set out by bus and were soon on the Peer Gynt highway. One stop was at a remarkable stave church, built in 1700. Our Norwegian guide, Arvie Mohn, was outstanding in pointing out features of the country and some of its folklore. Once, as we were going through a tunnel, he warned us against trolls, whose favorite food is American tourists. We were reminded that if one approached us we were to tell him, preferably in good Norwegian, that we had thrown a stone into the Arctic Ocean.

At one point we saw a lovely stream flowing under a bridge on a mountainside. Arvie had the bus stop while he told us the story of the "Billy Goats Gruff," which took place at that very bridge. His accent was wonderful, as the troll told the goats, "It is not allowded [*sic*] to go across this bridge." All along the way we saw lots of "cheep" [*sic*] and many "spurce" [*sic*] trees.

During one overnight stop we had a ride in a horse-drawn wagon to see a beautiful glacier. Several times we abandoned the bus for awhile and ventured by boat into one of the fjords. These narrow passages through tall cliffs often had waterfalls on their sides. Reflections in the clear water contributed to making these never-to-be-forgotten sights. We also did some amazing bus travel on highways from which snow had been blasted. Sometimes there were walls of snow higher than the bus on each side. Our driver, Bjorn, almost always had an apple to eat. He would drive with one hand and hold the apple in the other, but when he put the apple between his teeth and put both hands on the wheel we knew we were approaching a dangerous point. We were assured that all Norwegian drivers are good, because the bad ones are dead.

Our tour ended at Bergen just as midsummer was being celebrated by traditional bonfires. Another custom associated with that day is indulgence in the national drink, aquavit. This beverage must be bottled and then carried south of the equator and back on a ship before it can exert the potency with which it is associated. Those who believe in drinking it properly use a small glass with a round bottom which cannot be set down. This makes it necessary to drink the entire amount with one gulp. A little sip was enough for me.

Third Alexander Tour, 1977

The following year, David called his tour "Austria and the British Isles." Georgia could not go on this one, but I was happy to have my sister Louise go as my roommate. We first went to Copenhagen, and the next day flew to Munich and boarded a bus which was to take us through parts of Bavaria and Austria. Our first overnight stop was in Garmisch, where we marveled at the beautiful mountains and the many flowers. From there we went to Zell am Zee with stops on the way at Innsbruck and Obergammergau. There we visited the Christmas shop and looked at some of the work of the woodcarvers. Many of the houses had window boxes with red flowers, and there were also some with paintings of scenes from fairy tales on the outdoor walls. My favorite was "Little Red Riding Hood."

Zell am Zee, on a beautiful lake, is a lovely place to spend the night. We were entertained by some clever Tyrolian musicians in the evening. Salzburg was next, and there was a special stop on the way to see Golling Falls. This required a pleasant walk through a forest with lots of wildflowers, and was a refreshing interlude.

In Salzburg we visited Mozart's birthplace and the museum of articles relating to him. We saw a performance of his *Magic Flute* at a marionette theater. One of the outstanding places was the Maribel Palace, surrounded by statues and flower gardens. Another attraction of which Salzburg is proud is the site of the filming of *The Sound of Music.*

We had a stop at Melk Abbey on the way to Vienna. This is an old Baroque building on the Danube. One of its interesting features was that the paintings on the ceilings were shaped so as to give the impression that the ceilings were domed, when actually they were flat.

In Vienna the first place to see is the fantastic Schonbrunn Palace, chiefly a memorial to Maria Theresa, who ruled as queen or empress from 1740 to 1780. The boundaries of her territory varied from time to time as a result of wars and treaties, but it was centered in Austria. The palace was completed in the early years of her reign and the magnificent furnishings and decorations are said to be of her planning. Each of the many rooms had its own tasteful theme. In one alcove there were framed petit point pieces—one done by each of her ten children.

Another famous palace in Vienna is the Belvedere, built by a French general who was sent by Louis XIV to Austria to help the people get rid of the Turks—about 1698. The palace roof was built to resemble a Turkish tent, as a reminder that they would never again be invaded by Turks. Appropriately, the agreement which dictated the end of Russian occupation following World War II was signed in this palace. The Russians did not leave quietly, however. They had erected a monument in one of the city parks to their unknown soldier, and one of the conditions of their withdrawal was that this statue would be preserved. Our local guide told us that some attempts to surround it by tall trees or otherwise hide it had been made, but it was still very prominent in 1977.

A highlight of our stay in Vienna was to attend a performance of Mozart's *Abduction from the Seraglio* in the famous opera house. Louise and I were pleased to see that the name of the leading soprano was Moser—our maiden name. We had seen the name over several wine shops, and were glad to know that someone with that name had achieved status in another field.

A must in Vienna is a visit to the winemakers' village. We learned that people in Vienna drink wine–water is for bathing. If someone orders water in a restaurant he is the subject of pity, because he is very sick or very poor. The British Isles portion of this tour began in London. Tours of the city included several places I had not seen before, and our guide was entertaining as well as informative. In addition to serious history he knew stories such as why the Green Park was green. In earlier days there were beautiful flowers there and the King (I have forgotten which one) liked to walk there with ladies of the court and show them the flowers. The queen put a stop to this by ordering that all the flowers be dug up, and no plants with flowers have been planted there since.

We had a good bit of time for shopping and sightseeing on our own. I bought some dress material at Liberty and had a steak and kidney pie at Simpson's Restaurant. Delicious!

The bus tour from London provided a stop at Windsor Castle, and then again at Runnymede battlefield. From there we went to Winchester to see the cathedral built in 1070 which is the site of the tombs of King Canute and Isaac Walton. Next it was Salisbury and the cathedral with the tallest spire in England.

At Bath we saw some of the ancient Roman baths which had been unearthed and renovated. While we were having tea, our guide gave us the story (one version at least) of the origin of the four o'clock tea custom in England. During the early years of the British occupation of India, the men stationed there drank quinine water to ward off malaria. They discovered that this is much more palatable if mixed with gin, and so invented the first mixed drink, gin and tonic. They then decided to call everything to a halt at four o'clock and make a social affair out of drinking this medication. When the wives of some of the men went to India they wanted to be included in the party, but chose to drink tea. Biscuits and scones (pronounced "scuns") and other accompaniments were added later and the custom was carried back to England.

The real treat of this tour was to see Stonehenge. Although we had read about it and seen pictures, it was exciting to actually walk around the stones and hear about the various information which could be obtained from different view-  points. The theories as to who placed the stones and how they might have been brought there were interesting.

Stonehenge—England

We next visited Oxford, where we walked around Christchurch College and into some of the buildings. The dining room was especially interesting, because it had beautiful stained glass windows and portraits of some of the graduates, such as Lewis Carroll, William Pitt, John Wesley and Anthony Eden.

After stops at Stratford and Coventry we went on to Ruthin, in Wales. We spent a night in a 700-year-old castle, renovated of course. There were beautiful grounds with several peacocks, and a damp and gloomy dungeon.

Edinburgh was the next day's destination. While I had seen a good bit of the city when I was there on the study tour, we had not been able to go to Holyrood Palace because the Queen was in residence. Among the attractions in the palace were the bedroom of Lord Darnley, second husband of Mary, Queen of Scots, and "presum-

ably" (not my word, but that of our guide) the father of
James I. The room had a steep circular stairway leading
to Mary's room upstairs.

From Edinburgh we had a brief tour of Glasgow's
industrial district and then took a plane to Dublin, and
from there to Limerick. That evening we visited Durty
Mary's bar, then went next door to Bunratty Castle for a
medieval dinner. The attendants were wearing costumes
from the 1400s, and the food was presumably of the type
which would have been served then. A hunting knife or
the fingers were used for picking up the food. The mead
which they insisted that we drink was pretty awful, but
the medieval style music was good.

After a day of sightseeing and shopping in Limerick
and Killarney, we headed for Shannon Airport and home.

Delta Kappa Gamma Seminar in Stockholm, 1978

Although the "Seminar in Purposeful Living" is usually scheduled near the site of the convention, a bit of
divergence took place in 1978 when the seminar was held
in Stockholm and the convention in Chicago. Members in
Norway, Sweden and Finland were eager to be hostesses,
so we packed up and went to Stockholm.

Actually, Phebe and I had a travel agent plan a trip for
us which would take us first to Finland, then to
Stockholm, and finally to Denmark. We flew to Helsinki
and had a marvelous suite in an old but elegant hotel.
The elevator was called "hissi hiss." Thinly-sliced elk
steak was part of the appetizer at dinner. The special
Finnish rye bread was made in the shape of doughnuts
and served cold on a wire stand. The waiter asked us if
we liked it and Phebe said that it seemed a little hard.

Actually it was extremely hard. The waiter took a piece and broke it in two and said, "It's just right."

On a tour of the city we learned of some of the history of the country and saw several interesting old churches and parts of the castle fortifications which were still standing. Several places were closed because it was the eve of Midsummer's Day. We did get to visit the glass factory, where we learned their trick of getting a bubble of air into the stems of their goblets.

Birch Tree Branch
Finland's Symbol of
Midsummer's Day

Decorations for Midsummer's Day seemed to consist solely of having a branch of a birch tree. It didn't have to be of a special shape and did not have decorations—it just stood there, propped against a wall. We never learned of its significance except that it was a symbol of the day.

On the next day we had a boat ride across one of the many lakes to Tampere. At a revolving tower restaurant we had the traditional first day of midsummer meal, pea soup and pancakes with cloudberry jam. That evening we attended a planetarium show, listening to the lecture by translator earphones.

Our next destination was Jyvascyla, where Phebe had organized a Delta Kappa Gamma chapter several years previously. We joined a similarly routed group and were dinner guests of some local members. Some of the ladies wore traditional Finnish dresses, which were remarkably beautiful. We saw evidences of Russian influence as we walked around the town. The houses resembled those we had seen in rural Russia, and there was an Eastern Orthodox church.

In Stockholm we had time for a leisurely walk around the city. A kind gentleman told us we were walking on the wrong side of the street, and then suggested that we should go to see the changing of the guard at the palace. This was quite a show, with mounted guardsmen, a band, and some goose-stepping routines.

The formal opening of the seminar took place in the magnificent city hall. We were greeted by members of the host countries and served an elaborate smorgasbord. A lady who was a member of the city council brought greetings and there were remarks by other dignitaries. A tour of the building included seeing the hall in which some of the Nobel Prizes are presented.

The seminar program devoted one day to each of the host countries. Each one gave general information about the country's history, geography, music, and so on, with special emphasis upon the position of women and the educational system. Each country had planned a special midday dinner of native dishes. Norway's meal featured reindeer steak. The next day, Swedish members invited us to their opera house dining room where we had the greatest smorgasbord ever. There were such delicacies as raw eel, caviar, anchovies, all sorts of salads, goose, reindeer, steak tartare, veal and mushrooms, ham, and a great variety of cheeses and desserts.

Some of the United States members had arranged to take the group to visit the American embassy later that afternoon. We were greeted in a crowded little room by a woman in what can only be described as a "tacky" dress and given a few remarks of greeting by the deputy ambassador. Then we were served fruit punch—I think it was Kool-Aid—and some gingersnaps. I can't remember whether or not our drink was served in paper cups, but it would have fitted the occasion.

That evening the Swedish people sponsored a trip to Drottingham, site of the former summer palace. After a walk around the palace grounds we went to a theater built in the 1700s as the royal family's private theater. The performance was of Donizetti's *Elixir d'Amore*.

The ladies of Finland knew they could not top the elegance of the Swedish dinner, so they settled upon a typical farm menu consisting of a stew of meat and vegetables. Dessert was a custard topped by cloudberries. Incidentally these berries, which grow only in high-latitude countries, look like large colorless raspberries and are delicious.

The next portion of our trip began in Copenhagen, where we were met by a driver who was to take us along the Danish Riviera to Elsinore. We had insisted that the castle associated with *Hamlet* be included, and this seemed the only way. The moat had swans, but the almost-empty areas in the castle were somewhat dreary. It would be interesting to be there when a performance of *Hamlet* is in progress. The Fredenborg Palace, summer residence of the royal family, was much more beautiful. Back in Copenhagen our driver took us to several interesting places to which a tour bus does not go, and finally to our hotel. We shopped and spent some time in Tivoli.

The rest of the trip covered pretty much the same territory in Denmark which I had seen with the Alexander tour, but it was new to Phebe. We did hear some different stories, and it was a pleasant trip.

Christmas in Spain, 1978

The Delta Kappa Gamma 1978 Christmas tour to Spain found several good friends together–Mozelle, Rosalie, Georgia, Phebe and myself. We were delayed most of the night in New York, and landed in Madrid just in time for our tour of the city. I remember going to sleep standing up in the Prado, and the Royal Palace did not make much of an impression.

After a night of rest we were ready for the trip to Toledo, an ancient walled city on the Tagus River. We visited an impressive cathedral, but the chief attraction was the home of El Greco. We saw several of his paintings, including the famous *Death of Compte de Orgaz*. Back in Madrid we attended the midnight Christmas Eve service in the largest Catholic church in the city.

We were free to observe Christmas Day as we saw fit, and the hotel had planned for us a special lunch and dinner. One of the local guides told us of some of the Christmas customs in Spain.

The next day there was an all-day trip to the Valley of the Fallen, Segovia and El Escorial. The Valley of the Fallen has an enormous basilica built by Franco after the Spanish Revolution to honor those on both sides who died in the conflict. A cross on top of the building is 150 meters high, and it is surrounded by a group of huge statues. In fact, everything about this structure was big.

Segovia has reminders dating back to the Roman occupation several years B.C. We saw a plaque showing Romulus and Remus and the wolf, which had been sent to Segovia a few years previously from Rome. It was to commemorate the two-thousandth anniversary of the building of an aqueduct, which was still in use. Nearby there was an alcazar (fortified castle) in which the rulers of Castille had lived. Queen Isabella was crowned there.

On the way back to Madrid we visited El Escorial, a complex of buildings built to celebrate a victory over the French in the fifteenth century. It had an ornate castle and basilica, as well as art museums, a library and a college.

The next day we set out for Granada, passing by hills covered with olive trees. We went through la Mancha, the Don Quixote country. There were windmills now and then, but we suspected that they were for tourists to see rather than for use. Of course the main attraction in Granada was the Alhambra, built by Moors around A.D. 700 as a fortress and palace. We saw the luxurious Sultan's quarters and the extensive harem rooms. There were beautiful courtyards and gardens. At one passageway with a series of carved arches, our guide pointed out one arch which was slightly different from the others. He said that the Moors deliberately made errors now and then, to testify that "only Allah is perfect." On a hill across from the Alhambra there were numerous caves in which gypsies live. We were warned then and several times later of their proclivity for stealing, and their clever ways of doing it.

Another place of interest in Granada was the Carthusian Monastery, where we were told that the early monks there had only one meal a day and could speak only two words once a year. (I think the chances are that this was a joke of our guide, who had told us that 28 December is Innocents' Day, comparable to our April Fools' Day.) Currently the monks were making Chartreuse liqueur.

At one street intersection stood the statue, of which we had seen pictures in our history books, showing

Columbus kneeling in front of Queen Isabella as she gave him what our guide described as "the first credit card." Her crown, on display in the cathedral, had no jewels on it.

From Granada to Seville, at first we saw olive trees, grapes, tobacco, and vegetables, but as we went farther south there were oranges, cork oaks and palm trees. In one of the cathedrals we saw some of the 108 floats used in the annual Easter parade, which lasts for eleven hours. That evening we were entertained by flamenco dancers.

Among attractions in the city we saw the next day was the tobacco factory which inspired the opera *Carmen*. We visited the tomb of Columbus in one of the cathedrals. On the way back to Madrid we stopped in Cordoba to see a monstrous Moorish mosque–so big that a Catholic church had been built inside it. On the way into the city we saw a bridge which was built by the Romans and was still in use. The bus driver decided that the new one near it was safer. Our trip ended in Madrid.

Fourth Alexander Tour, 1980

An opportunity to attend a performance of the Oberammergau Passion play was the chief enticement of the 1980 Alexander tour for Georgia, Louise and me. On the way, our first stop was in the Netherlands. In Amsterdam we had a city tour which first showed us something of life along the canals. We then saw some of the famous paintings in the Reichsmuseum, visited a diamond merchant's workshop and sales room, and then had a look in the house of Anne Frank.

At Delft we watched the workers in various stages of producing the famous pottery, which we agreed should be called by a more elegant name when we saw the prices of the finished products.

From the Netherlands we went to Cologne and had a tour of the magnificent cathedral. Our guide told us of the excessive damage this building had suffered during World War II, and he was proud to say that this was the first structure to be repaired after the war.

Next we boarded a ship for a cruise down the Rhine. There were beautiful views of mountains and castles and pretty villages with lots of flowers. A speaker gave us information about the sights along the way, and the Lorelei song was played as we passed that point.

We left the ship near Weisbaden and went from there to Rothenburg, said to be the best preserved and most beautiful of the medieval cities. There is a story that years ago the mayor drank a gallon of wine in one gulp, on a bet, to save the city from invasion. This incident is reenacted every hour on the glockenspiel in the city hall tower. We visited the beautiful cathedral and walked a good distance on the wall which still surrounds the city after approximately five hundred years.

In keeping with a vow made in 1633 when the city was spared an epidemic of the plague, Oberammergau presents its Passion play every ten years. This endeavor, along with their woodcarvings and other attractions for tourists, occupies most of the time of the residents. The play is a day-long performance, with intermission at noon. It begins with the Palm Sunday entrance of Jesus into Jerusalem and ends with the Ascension. Every detail is skillfully included which tends to make it an authentic production. The casting, the acting, the costumes, and the choral renditions are simply marvelous.

Because the town is such an attraction for visitors, hotel facilities are crowded. We stayed in a private home which had been adapted for tourists. While the accommodations were not perfect, the house had boxes of red geraniums on each of its windowsills.

From Oberammergau we went through the tiny principality of Liechtenstein. Its location relative to mountains and air currents makes it a favorable place for hang gliders, and we saw several of these in flight. We did what tourists are supposed to do there—bought some stamps.

We were on our way to Lucerne for an overnight stop and some touring, and from there the next stop was Interlaken. The scenery along the way was magnificent, and we had a good view of the Jungfrau from our hotel window. We went from there to Montreux where we visited Chillon Castle, inspiration for Lord Byron's poem about the prisoner. We were now on Lake Geneva, and the next day we had a pleasant ride across the lake to the city of Geneva. Our tour of the city took us to the old League of Nations building, dedicated to Woodrow Wilson, and then to the United Nations building and the World Health Organization headquarters. There was also an interesting memorial to the Reformation, with statues of the leading figures involved in that movement.

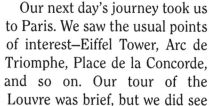

Arc de Triomphe—Paris

Our next day's journey took us to Paris. We saw the usual points of interest—Eiffel Tower, Arc de Triomphe, Place de la Concorde, and so on. Our tour of the Louvre was brief, but we did see *Venus*, the *Winged Victory, Mona Lisa,* and several other famous works. Then there was a more leisurely tour of the Cathedral of Notre Dame. It was all marvelous, and the rose windows startlingly beautiful. We also visited the Versailles Palace; it was even more magnificent than we had thought it could be. We had some free time for shopping and for a ride in a "bateau mouche" around the Isle de la Cité. This gave us different views of the city, including the back side of the cathedral.

From Paris we went to Mount St. Michel, with a brief diversion to Chartres. It was disappointing not to have time to visit the cathedral there. We climbed to the top of Mount St. Michel and heard about its origin and history.

The next destination was Calais and a ferry across the English Channel to Dover. The white cliffs are very impressive. We were delayed in getting a bus, so we had to make a nonstop drive to London.

Some of the sights in London were familiar, but this tour included St. Paul's Cathedral. The damage it had suffered in World War II had been repaired, and it looked clean and new. One chapel had been furnished as a memorial to American soldiers killed in the European Theater, and a book of 365 pages recorded all the known names. It was set so that one page is turned for viewing each day.

One evening we saw a performance of *Annie,* but the real treat was to see *The King and I,* with Yul Brunner in the role he made famous.

Greece, 1981

The Parthenon was one historic site which had always been high on the list of places I wanted to see. At its lofty position on the Acropolis it is the first thing one sees in Athens, and it was the first spot to which our tour of the city took us. While the Acropolis had been a center of activity since about 1100 B.C., the Parthenon was built during the Golden Age of Greece between 500 and 400 B.C. It was designed to house an enormous statue of Athena, goddess of wisdom and patron of the city. From the top of the Acropolis there is a view of much of the city, including the Agora, market place of ancient times, and the 2500-year-old Thission, best preserved of all the classical buildings. Mars Hill (the Areopagus), where justice was dispensed and where St. Paul preached, was pointed out to us.

During the rest of the tour and walks on our own, we saw the ruins of Hadrian's Arch, the temple of Zeus, and the oldest theater in Athens. At Sintagma Square, center of the present government, we watched the changing of the guard at the Tomb of the Unknown Soldier. The costume of the guards really does include the short pleated skirt and the pompons on the shoes.

We were in Athens a few days before the national elections and the four political parties each took a turn at having a nighttime rally on the Square. The communist

Temple of Poseidon
Cape Sunion, Greece

rally was to be held that day. We saw some of the preparations, but were advised not to go downtown that night. A beautiful drive took us to the top of Cape Sunion, a high cliff overlooking the very, very blue Aegean Sea. Ruins of the temple of Poseidon, god of the sea, stand majestically on this peak. The location of the temple was appropriate, because this is the last point sailors would see as they ventured off on an excursion, and the first sight of their homeland when they returned. Thus they had an opportunity to pray to Poseidon for a safe trip and to give offerings of thanks on their return. We were told that it was from there that Theseus, son of King Aegeus, said good-bye to his father as he set out to slay the Minotaur on Crete. On his victorious return, someone got the signals crossed and Aegeus, thinking that his son was dead, jumped off the cliff and drowned.

Our first trip outside of the area of Athens was to Kalambaka to see the Meteora monasteries, located on nearby high cliffs. These monasteries were built about five hundred years ago as sanctuaries for Eastern Orthodox monks. Only a few of the original ones still stand, and only five of those were occupied. While we reached the area by a steep winding road, the original inhabitants used a device consisting of a large rope basket which could let them up and down the cliff by a line on a pulley. Operation of this was demonstrated to us. There were beautiful gardens and attractive buildings, and the view of the countryside below was outstanding.

The next day we went to Delphi, passing Mt. Parnassus on the way. Delphi is the site of an ancient oracle and the ruins of the temple of Apollo. The oracle was located over a hole, from which steam arose, and was supposed to be the center–navel–of the earth. A priestess sat on a tripod over the hole and went into trances and made prophecies. The people had faith in these pronouncements since it was all under the control of Apollo. The best known story of the Delphic oracle is that it was there that Oedipus was told that he would kill his father and later marry his mother.

Next we went by ferry to Petra on the Peloponnesian Peninsula and then to Olympia. The site of the original Olympic games was naturally the attraction there. It all started as a religious shrine where Cronus was worshipped about 1000 B.C. Then Zeus (son of Cronus) came from Mt. Olympus and took over. The ruins of a large temple for worship of Zeus still stand. There was also a temple to Hera, wife of Zeus, and in front of that temple was the lamp in which the Olympic flame burned. The torch which carries the flame to light the lamp of present Olympic games is ignited there.

The games originated as the young men who came with their families to worship would run around the area and play for exercise. Games became an organized event first in 776 B.C. and reached their peak by 567, with performances every four years. Each state sent two athletes and thousands of spectators. Games and religious ceremonies lasted about a week, and all warfare between states was suspended for four months to give time for travel to and from Olympia. The stadium seated forty-five thousand, but this included only the men; women and children were not allowed to watch the games because athletes performed in the nude. (Of course, there were no female athletes.)

We saw what were said to be the original stones which marked the starting point of races. One of our group who was discus champion in his Senior Olympics had brought his discus and went out on the field to toss it. Another ran from one end of the stadium to the other. This was not pleasing to the guard in charge of the grounds.

When the Romans came during the second century B.C. they continued the games until A.D. 393, when

Theodosius I decreed that all pagan worship cease and Christianity be accepted.

In Nauplia, we stayed in an interesting hotel built on the side of a steep hill. We reached the lobby by going up an elevator located in a tunnel. We were there on voting day, and were left to our own devices as several members of our Greek travel staff had to go to their place of birth or their marriage to vote. This is a requirement of all Greek citizens.

The next day we drove to Epidaurus to see the only ancient Greek theater in which performances still take place. The stone seats are in a semicircular bowl-shaped area, and the enormous stage has marvelous acoustics. The main basis of plays still presented there is one of the three legends—Trojan War, Oedipus, or Ulysses.

The theater was built for the entertainment of people who came to worship and to seek healing from the god Aesculapias. He was the son of Apollo and a mortal woman who, having no husband, left the baby out to die. A shepherd found him and brought him up and taught him about the curative powers of various herbs. He became the god of medicine, and, among other devices, he had snake pits. We saw what was said to be the remnant of one of these. He used the venom for medications and mentally ill people were put in the pits for shock therapy. This is, of course, the origin of the two snakes on the caduceus—symbol of a physician.

The use of snakes actually explains why the Romans did not destroy the Epidaurus theater along with their other rampages. Word got to Rome that the venom from the snakes might be effective against a disease which was prevalent there. Someone in authority sent for some of the snakes, the treatment proved effective, and the area was spared.

On the way to Corinth we had a side trip to Mycenae, where we saw the remains of the Acropolis of the city and the famous Lion Gate. This area is thought to be one of the oldest settlements in Greece, as early as 3000 B.C. The legends of Agamemnon and the Trojan War originated here.

In Corinth, remains of the temple of Apollo still stand, but the Romans destroyed most of the area in order to make an agora. Attempts have been made to resurrect some of the Greek ruins without destroying the Roman ones by digging tunnels. On one Roman wall there is a note which says that St. Paul preached there. On the top of a nearby hill there are remains of the temple of Aphrodite, the goddess of love, which called forth some of St. Paul's strongest recriminations.

We crossed the Corinth Canal and went back to Athens and then to Piraeus to board the ship *Galaxy* for a three-day cruise of some of the Aegean islands. There was a brief stop at Mikonos and then we went to Rhodes. We did some sightseeing in the chief city, and then went to the other end of the island to Lindos. It is a beautiful town with a magnificent view of the sea. On one of the high points there is a monastery named Thenistes, dedicated to the Virgin Mary. Local belief is that if a woman has been married for a good while and has no children, she may walk barefoot to the monastery and spend the night. When she walks down the next day she will be pregnant. It must work, because we were told that there are several boys named Thenistos and girls named Thenista in the town.

Crete was the island of greatest interest. It is thought to be the site of one of the oldest civilizations and a step by which Egyptian culture was transmitted to Europe. Remains of the palace of King Minos have been partially excavated. It covered four acres and had twelve hundred rooms. There were openings now and then for air-conditioning, drains for rain water, and in the queen's quarters there was running water, a bathtub and a flush toilet. The worship of bulls probably gave rise to the legend of the Minotaur, which required that several young girls be sacrificed to it now and then. We had a guided tour through some of the labyrinth which was supposed to have led to the Minotaur's quarters.

Santorino, the next island we visited, had been the victim of a volcanic explosion in 1956. Half of the island had been blown away, leaving a high, barren wall. We were told that the only way to get to the town on the top of the mountain was by walking or by riding a mule. After a very unpleasant mule ride, we saw automobiles on the other side of the island, and roads leading up. We walked down.

Back to Piraeus and to Athens, we had two days of sightseeing on our own. Georgia and I spent a good bit of time visiting the site of the Stoa, where Socrates taught, as well as the agora and the marvelous museum where artifacts of each of the eras of Greek history were displayed.

✖ *Ring Around Central Europe, 1982*

After a lot of walking, looking and asking questions in the international terminal of Kennedy Airport, we finally learned that Yugoslavian Airlines does not have a counter until just before boarding time. Then a little strip with JAL on it is placed over a Pan Am sign. We found it, and took off on a very pleasant flight to a place with the unpronounceable name of Ljubljana. From there a bus took us to a lovely resort town named Bled, on Lake Bled. After a walk around to look over the area, we had a nice dinner which featured baked trout and paprika salad. (Just chop up a lot of paprika peppers, add a little dressing, put it on some lettuce, and you have that salad—sort of hot, but good.)

The next day we visited a castle which was originally built in the 900s by the Celts. Since then it has been rebuilt and modified by Slovenes, Romans, Croats, Turks, Austro-Hungarians, Germans and possibly others. There is a separate room furnished to represent each of the periods of occupation. We also saw a beautiful old church and then went to a gorge with lovely waterfalls and interesting ferns, mosses, and liverworts.

The next day we crossed the Loibl Pass to Austria, passing through areas of magnificent mountain scenery. As we neared Vienna the land became level. Our hotel reservation had been changed, and we were sent to a small place a long way from the center of the city. We were hungry, and there was no dining room. The only place we could find open was, of all things, McDonald's.

On our tour of the area we went to a cavernous and gloomy gypsum mine, which had been used during World War II as a secret place to make airplane fuselages. A finished one was taken out at night and the wings, made someplace else, were attached. Other sightseeing included several places I had seen previously, but this time we visited the Spanish Riding School and watched the Lipizzaner horses go through some of their routines. In the evening we were entertained in a park by Viennese music and dancing. Georgia and I figured out how to use the subway and made some other excursions into the city center on our own.

We had a hydrofoil ride down the Danube to Budapest. Our hotel was on the Buda side of the river. A typical Hungarian dinner featured goulash, while we were entertained by musicians in native dress. It was a very lively evening.

One of the impressive sights on our city tour was the Cathedral of St. Stephen, the first King of Hungary. We heard something about his life and learned that he had been designated a saint because of his enthusiasm for converting the people to Christianity. Another site of note was the Hero's Monument in the main city square. The central part of

Section of Fisherman's Bastion—Budapest

the impressive monument was of men who represented the six tribes which originally united to form the Hungarian nation. Many other statues representing later phases of the country's history surrounded the central figures. The impressive House of Parliament was also on our agenda. We learned that the very elegant lower house chamber was not in use, because the body for which it was intended did not exist. At a high point above the Danube there was a statue commemorating the Russians who died in World War II.

From Budapest we had a flight to Belgrade, capital of Yugoslavia, located at the point where the Sana River joins the Danube. Its location has made it an important site for fortification throughout the centuries. On a high point above the junction of the rivers there were ruins of walls built, in turn, by the Romans, the Turks and the Austrians. We were told that Belgrade has been ruled by sixty-five different conquerors or liberators. At the time we were there, Yugoslavia was made up of six states with different backgrounds and languages represented, held together as a Communist republic. We visited the tomb of Tito, the honored leader who had played an important role in the more recent history. It was a beautiful tomb surrounded by many colorful flowers. A museum of Tito's belongings was nearby, but we did not have time to visit it. We were told, however, that one of the mementos was a rock from the moon, given to Tito by President Nixon.

Our next stop was at a village near Dubrovnik. Our hotel, the Croatia, was built on the side of a cliff overlooking the Adriatic Sea. This is a popular resort area, with terraces in the side of the cliff where people engage in sunbathing. Georgia looked out the window and said, "They all have on the same color bathing suit." Then, "No, they don't have on bathing suits."

Dubrovnik is on a strategic location on the Dalmatian Coast. Its walls, built about 1300, still stand (or did when we were there). Half of the city was Roman and half of it was Slavic, and a wide street separated the two parts. We visited several of the interesting buildings and strolled along the wall to see the city up on a hill, and the Adriatic Sea below.

Back at the hotel we learned that we would leave the next day for Sveti Stefan, as our tour director was not satisfied with the service we were getting. We had a last look at the sunbathers, and then saw a beautiful sunset and later a new moon, over the Adriatic.

Our ride to Sveti Stefan (St. Stephen in another Slavic language) was southward along the narrow strip of Croatia which extends almost to Albania. There were vineyards and olive trees and vegetables and fruits. Our local guide told us that most of the farmers were independent, and only twenty percent belonged to communes. According to him, only about twenty percent of the people belonged to the Communist Party, but it was the only one. Farther along the way we got into mountains, and crossed a fjord on a ferry.

Sveti Stefan was an old walled city on an island. A causeway has been built to connect it to the mainland, and the buildings have been converted into a resort area with many individual cottages. It was a lovely place to be.

We went back to Dubrovnik to get a flight to Zagreb, capital of the province of Croatia. During free time the next morning, I went on a walk through the botanical garden. It was not very well kept, and when I saw a sign which identified Canadian hemlock as native to South America I decided that it needed lots of attention. Zagreb is a very old city with a building which was the first University of Yugoslavia, built in the 1000s, and now being used as an elementary school. We saw a St. Stephen's cathedral and a St. Mark's church on the roof of which the coat of arms of Croatia and of Zagreb are designed in colored tiles. There was a Tito Square with beautiful plantings.

We said good-bye to Zagreb and went to Belgrade for our flight home.

✠ *Iberia, 1982*

Phebe and I decided that we did not get a fair chance at seeing the Prado in 1978, so when we saw that Caravan Tours would take us to Portugal, Spain and Tangier in November of 1982, we were on our way. Our first stop was in Lisbon, an interesting city with sidewalks paved with black and white tiles. On our tour of the city we saw several churches and castles which featured the Manueline style of architecture, identified by its use of carved ropes. One special sight was a carriage museum, the only one in the world, we were told. It had been customary that each sovereign have a carriage designed especially for him, and a number of these had been preserved. They were very elaborate and heavy; at least six horses, and sometimes as many as twelve, were required to pull one of them.

On our way to Sintra we crossed the Tagus River on the longest bridge in Europe. At Sintra we visited the eighteenth century Queluz Palace, patterned somewhat after the one in Versailles. Walls were covered with silk, and there were beautiful rugs and paintings. It was very rainy and windy, and high waves were beating against the sea wall with such force that spray came across the road, so we had to go back to Lisbon before we finished the tour. Back in Lisbon we went shopping in a beautiful mall across the street from our hotel and were almost blown away when we tried to get back.

Everything was calmer the next day when we set out for the fishing village of Nazare. Waves were still high, however, and boats had been taken into shore for protection.

The Shrine of Fatima had its origin on 13 May 1917, when three children had a vision of the Virgin Mary. A large church had been built there and many thousands of pilgrims come to worship and to partake of the healing waters of a spring which spurted up after the vision.

On the way to Spain we stopped at Coimbra University, founded in the

Branch and Acorn
Cork Oak Tree—Portugal

105

eleventh century, and of course the oldest in Portugal. Along the way there were olive groves, vineyards, and lots of cork oak trees. Cork products, we understand, make up forty percent of Portugal's exports. Farther on there were cattle and sheep, and as the land became flatter, fields of wheat and corn. In Spain we had a stop at Avila, the walled city which I remember from a picture in a history book. The walls built in 1090 were still standing, and we were allowed to climb at one place and take pictures.

Next we stopped at the Valley of the Fallen, which we had seen previously. One thing we were told this time is that if one of the crossbars of the cross were in a proper position, two buses could pass on it. When we got to Madrid we went first to El Prado on our own, and later with the group. In the Royal Palace I had not remembered that there are three hundred clocks in the building, some of them very elaborate. Six men are employed full-time to wind and care for these clocks. The palace is also famous for its many beautiful chandeliers, and for a dining table which will seat 140 people.

We went again to Toledo, Cordoba and Granada. One new story about the Alhambra was that the sultan had blind musicians play for the dancing of his concubines.

As we went toward the Mediterranean we stopped at Myas, whose claim to fame seems to be that it had the only square bull ring. I chose not to go to the bullfight, but Phebe went and said that it was not terribly gory. Our destination was Algeciras, where we boarded a ferry across the Strait of Gibraltar to Morocco. Of course I took pictures of the Rock, but never got a view like the Prudential logo. We landed at Ceuta, which is owned by Spain, then went by bus through the Riff Mountains to Tangier. Our guide, Ahmed, advised us that there were four things worth buying in Tangier: rugs, brass, copper and leather. All the rest, he said, was "hunky-junky." He took us to the Casbah, formerly the king's palace, built somewhat like the Alhambra. The open courtyard was very crowded, and we were besieged by people wanting to sell us something and by children begging for "bonbon."

After our dinner of couscous, we were entertained by music and by belly dancers. The next day we went back to Algeciras and then to Seville. On the way we were treated to a fantastic sunset. The entire sky became a brilliant red. In Seville we saw some different sights, including the third largest cathedral in the world. Previously we had simply seen the chapel in which the tomb of Columbus is located.

From Seville we went back to Lisbon, crossing on the way a bridge over the end of which was an enormous statue of Christ. We learned that this statue had been given by the women of Portugal as an expression of thanks that they were not involved in World War II. From Lisbon we returned home.

✖ *Ireland, 1983*

I had seen a little bit of Ireland on the Alexander tour in 1977—enough to make me want to see more. Georgia had never been there, but said she had wanted to go for some time because she had married the first Irishman she ever met. With these incentives we decided in 1983 to take a tour which would give us two weeks in Ireland. After landing in Shannon we did what seems to be a requirement of all tours to Ireland: the Bunratty Castle medieval banquet. The next day we headed for Galway, with a stop at the Cliffs of Moher. This is a series of cliffs—some as much as seven hundred feet tall—of dark stone

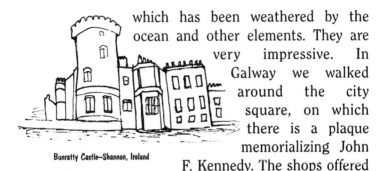

which has been weathered by the ocean and other elements. They are very impressive. In Galway we walked around the city square, on which there is a plaque memorializing John F. Kennedy. The shops offered

Bunratty Castle–Shannon, Ireland

tempting items for purchase—linens, woolen goods, china and crystal.

The Aran Islands stretch across the entrance to Galway Bay from the Atlantic, and have such interesting names as Inishmore (the largest), Inishmaan and Inisheer. We had a flight to Inishmore on a five-passenger plane for which we had to be weighed so that the load could be properly adjusted. Transportation on the island was by jaunting cart. We did not get very far before it started to rain, so we did not get to see all of the scheduled places. We found a shelter in which to eat our box lunch, then returned to the "airport" in what became torrential rain, along with strong wind. There was some uncertainty for a while as to whether or not we would be able to return to the mainland. The weather did improve and, soaking wet and cold, we got back safely.

Later, after dinner, we were entertained by music of accordion, harp, piano and whistle—not all at the same time. The master of ceremonies interspersed jokes between the musical numbers. One example concerned a school in which an Irish teacher, on St. Patrick's Day, told her pupils that she would give a dollar to anyone who could name the most important saint. The English boy said "St. George," the Scottish boy said "St. Andrew" and the Jewish boy said "St. Patrick." As the teacher gave him the dollar, she expressed surprise at his answer. He replied, "I know in my heart it is Moses, but a deal's a deal."

On the way from Galway, north to Westport, we visited the Connemara marble works and watched some workmen shaping various articles from the marble. It came in several shades—white, dark and light gray, and green. Another stop was at a peat bog, and we could see where blocks of peat had been removed. We were told that there are commercial agencies which use machines, but that many people cut out their own supplies with a special kind of shovel. After the peat has been properly dried it will not absorb water. It is, of course, a major source of fuel in Ireland and there seems to be plenty of it, although it is estimated that one thousand years would be required to turn the present vegetation into peat. At Westport we were entertained by music and dancers of the Irish jig.

Our Donegal excursion took us farther north and included crossing on a swivel bridge to the island of Achill. Views of the cliffs were magnificent. Back on the mainland we went through an area with French influence, due to the fact that Napoleon had sent some of his soldiers there to help the Irish fight the English. We came very near the border of Northern Ireland at one point. Our guide told us something of the dissension between the Republic and the North area, and concluded by saying that it is "a struggle between two religions with no Christianity."

On past Donegal we went to the most northwestern point in Ireland, called the "Bloody Foreland." There was evidence along the way of glacial activity in an earlier age. We returned to Donegal for a look around the town and then went to Bundoran, on Donegal Bay, for the night.

Next we headed south to Dublin, with a stop at the cemetery in which Yeats was buried, at Tara, and at several other historic sites along the way. Only mounds and markers identify the Great Hall of Tara and the palaces and other buildings which, from early time to the sixteenth century, constituted the seat of Irish kings.

The high point of our tour of Dublin was the Trinity College Library. The beautiful building with a cathedral-like ceiling had walls stacked with rare books, as well as glass cases with displays. The outstanding one was the *Book of Kells*, said to be the most beautiful book in the world. It is a Latin translation of the four Gospels, written in beautiful script and ornamented by colorful illuminated letters and pictures.

One evening in Dublin we attended a production of Shaw's *Candida*. It was quite interesting. Jokes and music were our entertainment the next evening. Sample joke: A fellow was weaving along the highway and a policeman stopped him. The policeman said, "Do you know you are drunk?" The fellow answered, "Thank goodness! I thought my steering was out."

From Dublin we went to Wexford, an old town dating back to 300 B.C. We had a walking tour conducted by a stately old gentleman, a member of the Old Wexford Society. He told us that the area was first settled by Danes, and that the Normans came in 1169. This lasted until 1921, when all of Ireland except the northern segment became free of England.

Our next stop was Waterford, where we saw a film (Irish pronunciation "fillum") showing the making of the famous crystal. We also visited the showroom with many special single pieces as well as samples of the various patterns. They did not have a sales room, but several shops in the town offered many lovely pieces for sale. On the way to Blarney we had lunch at Cork, center of a prosperous-looking manufacturing area. Of course everyone is expected to kiss the Blarney Stone, which is supposed to give the gift of eloquence. The stone is located in Blarney Castle in a place which makes it necessary to lean over backwards to get in a position to kiss it. I haven't recognized any improvement in my ability to speak, but the castle grounds were beautiful.

Killarney was our next stop, and we enjoyed a jaunting cart ride through the beautiful lake country. Our bus then took us around the Ring of Kerry, justifiably noted for its magnificent scenic views. From Killarney we went on to Shannon for our flight home.

Middle Europe, 1985

This trip in 1985 covered some of the same territory as "Ring Around Central Europe," but it also took us to Poland, Czechoslovakia, East Germany and West Berlin. We landed in Belgrade, and after an afternoon of rest we were entertained by lively folk music and dancing. During our dinner an artist came to the table and made a sketch of one of our group. A tour the next day took us again to the strategic point above the confluence of the Sana and Danube rivers. In addition to a review of the history of the area, we heard a story about a statue which stood at a point overlooking the city. This statue of a nude young man had been designed to be placed on a pedestal in one of the city squares. Mothers of the city protested, saying they did not think it proper for their children to see the figure of a man "dressed only in nature." So the statue was removed to the high cliff, too far away to be seen clearly from the city.

From Belgrade to Budapest we had views of fertile land—crops mainly of corn and sugar beets—and a number of small villages. After a guided tour of Budapest, Georgia and I had a leisurely walk around the city, then spent some time in the Museum of Fine Arts. There was an exhibit of Egyptian art, and a section devoted to interesting old etchings.

The next day we had a bus ride through Czechoslovakia on our way to Krakow. There were many interesting-looking villages, often with onion-domed churches which were reminders of the Turkish occupation. We had a lunch stop somewhere in the Tatra Mountains, which were beautifully lined with symmetrical fir trees.

Our guide in Poland, Andy, gave us a rundown of our program, and charged us $43.00 for extras. He also offered to get Polish money for us from the black market. A special city guide in Krakow told us that it had supposedly been made into a model communist city, but that the people did not like it because they had to work harder for less money. They had especially felt the need for a church, and after several forms of protest they were given permission to build one. Everybody turned out to build the church, which is covered on the exterior by pebbles collected from nearby streams and put in place by the collector. Building a second church had been accompanied by dissension and riotous incidents also, and during that time a priest had been murdered. Pope John Paul II had visited the unfinished church not long before we were there, and enlarged photographs of him and the enormous crowds which came to see him were prominently displayed.

Krakow had been the headquarters for Hitler's army during World War II and the Russians reportedly drove them out so quickly that they did not have time to blow up everything. Hence it is the only major city in Poland which has antiquities. Among other places, an example we saw of this was a palace built in the 1300s, with beautiful tapestries, furniture and other works of art.

We visited a nearby salt mine, walking down eight hundred steps to the bottom. There were many corridors, and numerous figures carved from salt. There was even a salt chapel, complete with altar, chairs and chandeliers. An elevator brought us back to the surface.

From Krakow we headed toward Warsaw, with two stops along the way. The first one was Auschwitz, where we were given a tour of a German concentration camp. It was too horrible to think about, as it confirmed and added to what we had heard about the inhuman activities that went on in those places. We were amazed to see groups of school children in one of the buildings, but were told that they must see it and resolve that they would never let anything like this happen again.

Our second stop was at Czestochowa for lunch and a view of the *Black Madonna*. This is a painting which was the only object not destroyed in a disastrous fire some years ago. At 3:30 a choir began to sing, several priests appeared, and the curtains in front of the picture were opened. We were told that on certain days related to the painting, thousands of people come on foot, often from many miles away, to worship at this shrine.

Our tour of Warsaw took us to the ultramodern memorial to Chopin. The original one was destroyed during World War II, but the model was later found in the home of the son of the sculptor, and a replica was made. We were told that during the German occupation anyone caught playing Chopin's music was sent to a concentration camp.

We saw a film showing first the city as it had been before the war, then moving to some pictures made by the Germans showing their systematic destruction of everything in the city, and finally views of the restoration. By using pictures, and what remained of the buildings and their foundations, the rebuilt structures had been made to look as much like the original ones as possible. A new building, similar to several in Moscow and the tallest one in the city, had been given to Warsaw by Stalin. Andy told us that from the top of that building you could get the most beautiful view of the city—because when you were on top of that building you could not see it.

From Warsaw, our next overnight stop was at Poznan, near the border of East Germany. On the way we visited the birthplace of Chopin, Zelazowa Wola. In the house there were mementos of Chopin, including a piano on which he had played. The grounds were beautiful and well-kept. The shops in Poznan featured jewelry and other items made of amber. The chief source of this material in the world is in the area of Poznan, near the Baltic Sea.

At the border we had to wait about an hour at a holding station before we were allowed to cross the Oder River and enter East Germany. We arrived shortly at East Berlin and had a guided tour of the city. We rode on "Unter den Linden," with young linden trees having replaced those destroyed in World War II, saw the

Brandenburg Gate and several government and business buildings, and so on. The greatest amount of time was spent in the Pergamon Museum, with its extensive displays of archeological materials from areas of Turkey and Iraq–some dating back to 135 B.C.

In our hotel room Georgia and I decided we would like some ice. We knew that "eis" means ice in German and that it also means ice cream. Georgia very carefully explained to the room service person on the telephone that we wanted ice and not ice cream. After a while a waiter arrived with a very large and beautiful chocolate and strawberry sundae. We gave up and decided just to pay for it, but the price was 900 marks, and we had only dollars which were not acceptable. The waiter agreed to let us sign for it, and left. Very soon the boss came, but he changed $3.00 for 900 marks and all was well. Meanwhile, even though it was near dinner time, we ate the sundae. When we told about this at dinner, we were informed that there was a free ice machine near the elevator on the hall.

Our next venture was to West Berlin. Our guide told us that she could not go with us because West Berlin would not accept her. Of course we knew quite well that the real reason was that East Germany would not allow her to go. We had a rather nervous wait at the Wall, while a stern-looking military man wearing several guns took our passports for inspection. At last we were allowed to go through Checkpoint Charlie for a brief visit in West Berlin. We saw the mound where the building in which Hitler is said to have committed suicide once stood, as well as the Reichstag, Charlottenburg Palace, a number of statues and memorials, and the stadium of the 1936 Olympics. Then we returned to East Berlin for the night.

From East Berlin we stopped first at Dresden, having heard along the way a lecture from our guide about the great advantages of living in communist East Germany. In Dresden we were told, with a hint of resentment against Americans, about the bombing of the city during the war. Most of the damage had been repaired, but a portion of one bombed building had been left as a memorial. In the Dresden Porcelain works we saw a number of exceptional and rare pieces on display.

This time we crossed Czechoslovakia nearer the western end and had a one-night stop and a day of sightseeing in Prague. At the old castle which is located on the highest point of the city, we watched the changing of the guard, and then went to the Cathedral of St. Vitus. We came down to the city square and then had a stop at the Jewish cemetery. In this unusual cemetery graves had been dug on top of graves and the whole thing was a jumble of unorganized markers. I was especially pleased then to get a visit to the Moser Crystal showrooms and sales area. It was too expensive to buy much, but I just had to buy a few pieces. It is quite elegant.

We had become somewhat familiar with Vienna, so when we got there we did our sightseeing on our own, and enjoyed it very much. We came across the Moser Gift Shop, and prices there were not so steep. From Vienna we went back to Yugoslavia for our flight home from Zagrab.

Sorrento and Rome, 1985

This trip, taken later in 1985, was different from previous ones. We stayed in the same city for about two weeks, taking optional tours to surrounding areas of interest. Sorrento seemed ideal for this sort of venture and a few days in Rome were an added attraction. Our hotel room had a balcony from which we could see the Bay of Naples and Mt. Vesuvius. We were within walking distance of the center of the city, and buses were available.

The first excursion was to Amalfi. The road crossed mountains of the Salerno Peninsula to the Gulf of Salerno. Views were beautiful, but a little scary when the bus had to back up and get a fresh start in order to get around some of the steep hairpin curves. Along the coast of the gulf there were vineyards and olive orchards and several neat little towns tucked into the cliffs.

Buses were not allowed to go into Amalfi, so we shifted to water transportation. At first a small boat took us into the Green Grotto, a limestone cave with strange formations and an eerie green glow on the water. Along the shore in a larger boat we saw beautiful hotels and houses. The home of Sophia Loren, which had been confiscated because she didn't pay the taxes, was pointed out to us. We had a grand view of Amalfi from the boat, then walked up to the city for lunch and a brief tour. It is truly a beautiful city in a fabulous location.

On one of our tours we drove around the area of Sorrento itself to see the marvelous views of the mountains and the bay. In the city we went to a mozzarella cheese factory. We saw the entire process, from warm milk just extracted from the cows to the finished product. One of the last stages reminded us of a taffy-pulling. Two men took a batch of the material and pulled it, then folded it over, repeating many times until it reached the right consistency. We understood why mozzarella tends to be stringy sometimes. We also visited a winery, then an inlaid wood factory. All sorts of objects decorated with designs of the inlaid wood were available.

A trip to Monte Cassini made a stop at Casserta, site of what is claimed to be the largest palace in the world— one thousand rooms. It was built in the 1700s for the King of Naples. The Allies used the palace as headquarters during World War II, and parts of it were still under repair. What we saw of it was very elegant.

Monte Cassini is the site of a Benedictine abbey, founded in the first century. In its early years it was influential in spreading Christianity. It is strategically located on top of the mountain, and was selected for the headquarters of Hitler's army. Many thousands of Allied soldiers were killed in an attempt to dislodge the Germans. Finally a decision was made to bomb them out. A cemetery for Polish soldiers killed in the battles there was visible from the mountaintop. Later on we went to the nearby British cemetery of over four thousand graves. American soldiers killed there were buried at Anzio. The Abbey and other buildings have been rebuilt by contributions from the Allied nations.

On the tour to Vesuvius we stopped in the present-day town of Pompeii to see the artisans carving cameos. It was fascinating to see how they use the various layers and colors of seashells to make the designs. Some of the display items were very large and elaborate.

The bus went as far as it could up the side of Vesuvius, but there was still a long climb through loose granular soil to the crater. That is a big, big, deep, deep hole! From

Vesuvius

the edge of the crater we could look around and see various levels of lava deposit indicating the more recent eruptions.

The disastrous eruption in A.D. 79 destroyed the cities of Pompeii and Herculaneum. Evidence indicated that Pompeii, nearer the volcano, was destroyed without much warning by hot ashes, dust and gases. Herculaneum evidently was covered by rivers of mud, and it is assumed that many people had time to escape. Both cities remained buried until they were somehow discovered– Herculaneum in 1709 and Pompeii in 1748. A lot of digging throughout the years had exposed probably most of Pompeii. Herculaneum, under solidified mud in some places one hundred feet deep, had a town built on top of it. The point to which excavation had gone is a high cliff with houses and other buildings, which cannot be arbitrarily removed, on top of it. It is a strange sight.

We had a rather extensive tour of each of the cities. In Pompeii there were stone buildings, walls and paved streets. Columns marked a place which must have been the forum. There were shops with indications of the kind of business carried on, and homes with mosaic floors, and paintings on the walls. The shape and position of human bodies had been obtained by pouring cement into cavities left by decayed bodies, allowing the cement to harden, then removing the softer surrounding material.

Herculaneum had not been subjected to burning, and organic materials such as wood had been at least partially fossilized. Buildings, furniture, and other household articles had been well preserved. In what was apparently a weaver's shop, a loom stood unbroken. There were tiled floors, paintings on walls, and other indications of lifestyle prior to A.D. 79. We learned that many valuable articles from each city had been taken to a museum in Naples. This was, of course, one of the attractions for a tour of that city.

Naples is a beautiful city with wide streets and a number of attractive parks and squares. It was at one time a kingdom and the king's palace still stands. There is a well-known art museum, but our interest was chiefly in the archeological museum where treasures from Pompeii and Herculaneum were on display. A number of complete walls with frescoes on them had been brought here for safety. There were mosaics, furniture, jewelry, statues, and many household articles.

In the astrological section of the museum there was a very interesting display. A picture of each of the signs of the zodiac with the months and days it represents was arranged in a special way on the floor. A hole in the corner of the room lets in rays of the sun, and all year long on every sunny day at noon a spot of sunlight shines on the exact date of the proper month.

Out in the suburbs we were taken to a "dying" volcano which was giving off puffs of steam and sulfurous vapor. When the attendant touched a piece of burning paper to the ground, all of the puffs increased. Interested as I was, I must have walked too close to some of the little wisps. When I got to the hotel room and took off my hose, I realized that they were full of tiny holes, which turned into a mass of runs.

Having come along when "Isle of Capri" was a popular song, I thought of that island as a romantic place. I'm sure that under certain circumstances it could be romantic, because it is very beautiful. A little boat took us into the Blue Grotto, much like the green one in Amalfi, but somehow allowing different wavelengths of light to come in. At the top of the island we saw ruins of an ancient palace of Tiberius, and many lovely homes and gardens. Two very high stone pillars at one end of the island were said to have served as "lighthouses" in ancient times. Each evening slaves would climb up these rocks and light a fire on the top of each one. They made use of what they had!

One of the tours was to Paestum to see the ruins of three temples built by the Greeks during their occupation around 550 B.C. These temples had been overgrown by trees and other vegetation for about two thousand years until they were uncovered in the 1800s. Although this was an area of earthquakes, many columns of the temples were still standing. One possible explanation of this is that the temples were built on sand, which allowed the ground to shift rather than crack open during a quake. One of the temples had an outer area evidently used for sacrificing animals, because there was a pit nearby with bones and other residue. After the Greeks left, the Romans had apparently used the site for a forum, and had built a theater. (As all students of archeology know, Greek theaters were half-circles and Roman theaters were complete circles.)

During our stay in Sorrento, Georgia and I walked several times to the city for shopping and looking. We were always intrigued by the colored nets stretched under the olive trees to keep the olives from falling to the ground. At the hotel there were lectures in the evening. One of these was on the Italian language, one on the Italian educational system, and one explained the process of making the inlaid wood designs. We enjoyed sitting on our balcony looking at the beautiful view, but we had to leave it and go to Rome.

We were still with the tour group in Rome, and had several tours with them—one to St. Peter's and the Sistine Chapel, one to the Colosseum, and one to a museum on Capitoline Hill. In this museum there were many paintings and tapestries, and among the statues were the one of the *Dying Gaul* and the bronze figure of the wolf and babies, *Romulus and Remus*. Our hotel was near enough to the center of things for us to walk around the city and sightsee on our own. On Sunday we took a bus back to the Vatican and happened to get there for a special mass in St. Peter's. One group after another of colorfully robed bishops, cardinals and priests came in, and then a large choir in white robes. We didn't find out what the occasion was, and since it was so crowded that we had to stand, we did not stay long. We went out into the plaza and waited, along with mobs of others, for the Pope to come to his window and bless us.

❈ *Garmisch-Partenkirchen, 1987*

This was a week-long trip in November of 1987, during which we again spent most of the time in one city, with optional tours to nearby places. Garmisch-Partenkirchen is itself a lovely place in which to spend a week, and we enjoyed walking around and getting acquainted with the city. Among the places we went with the tour group were Innsbruck, Oberammergau, and Neuschwanshein Castle. We organized our own tour to the top of the Zugspitz. The last day was spent in Munich.

We had been to Innsbruck before, but it is beautiful at any time. On the way we had a stop at the Swarovski show rooms, with rows and rows of beautiful carved crystal, and lighting arranged so as to make it really sparkle.

Some of the pieces were truly fantastic. In Oberammergau we had a tour of the Passion play theater, including a look at the costume and property rooms. Neuschwanshein Castle, near Fussen, was built by Mad King Ludwig about 1875. It is, of course, the castle pictured in Walt Disney promotions. It is outstandingly beautiful from any angle, as well as inside.

We had a wonderful trip by ski lift to the top of Mount Zugspitz. Georgia had been there before and had seen nothing but fog, but we were lucky to have a clear day for the view on the way up as well as from the top. It was extremely cold and windy, but we got some good pictures of the peak. Then we went into the building near the top, where there were charts and information about the mountain, as well as warm beverages.

The shops in Munich were closed because of a religious holiday, but Hofbrauhaus was open. We joined the merriment there for a while, and then went to the airport for our flight home.

✖ Ireland, 1988

It can safely be said that Georgia is a devotee of the United States Military Academy at West Point, having had a husband and four sons graduate from that institution. When she received word that the annual football game between West Point and Boston College was to take place in Dublin in 1988 she felt that she must go. It was easy for her to persuade me to go along, although I said at the time that if the game were being played in Fayetteville, I probably would not go. The two institutions had been in the process of planning this event, we understood, for three years.

The game was played in the largest soccer stadium in Dublin, and a great many natives joined the thousands of Boston and West Point fans, who had come by chartered planes, to fill the stadium to capacity. Both institutions had brought their bands. West Point had its plane and parachute jumpers ready for the traditional presentation of the game ball. They could not, however, match the appearance of the Boston College women cheerleaders. On the way back to our hotel after the game our driver told us that he didn't know much about the game, but that he surely enjoyed the cheerleaders. I believe that West Point had been favored to win the game, but it didn't come out that way. I don't believe it dampened their spirits very much, however.

We had a bit of sightseeing across the country from Shannon Airport and back, but I don't recall that there was anything special that we had not seen before. We spent a night in Galway. The weather was nice and I thought I was going to see the sun go down on Galway Bay, but some clouds came in a few minutes too soon.

Canterbury, 1989

The Delta Kappa Gamma chapters in Great Britain have been assigned to the Southeast region of the society, and in 1989 the members there invited the United States members of the region to attend their convention in Canterbury. Phebe and I went, along with about two hundred other members from this country. We had rooms in a dormitory of the University of Kent at Canterbury.

It was interesting to attend their workshops and meetings and to observe how Delta Kappa Gamma methods and goals operate in another country. Their banquet was somewhat different from ours. The Right Worshipful Lord Mayor of Canterbury attended, wearing his elegant red robes and other regalia of office. Prior to the dinner, we had toasts to the Queen, the Mayor, a number of other special guests, and several Delta Kappa Gamma officials. Each person giving a toast was introduced by the toastmistress with the phrase, "Pray silence for . . . "

Naturally we did as much sightseeing in the city as possible. The interesting old cathedral was the most important site and we stood where many pilgrims have stood—at the place where Thomas Beckett was murdered. We also found out about a place where *The Canterbury Tales* are relived. It has been developed in a series of underground passages which lead from one to another of several places designed to represent the overnight stops of Chaucer's pilgrims. A recorded speaker reports progress between stops, and each inn is an authentic-looking setting, with life-sized models of some of the pilgrims. The speaking system relates, using a different voice for each character, an imagined conversation between the other pilgrims as they react to the tale of the day. It was very well done.

Greece, Turkey, Israel, Egypt, 1989

Phebe had occasionally said that although other responsibilities prevented her from going often on a long trip, she did want to go to Greece some day. That day came when we were able to make arrangements for a combination tour of Greece, a cruise of some of the Greek islands, and an additional cruise with stops at Turkey, Israel and Egypt.

The land part of the tour was in two segments, before and between the two cruises. It covered essentially the

same places I had been before, but I learned something
new at each place. The same islands were also visited, but
this trip included Delos as well. This island was sacred to
the ancient Greeks because it was the legendary birthplace of Apollo and Artemis. The outstanding sites were the Terrace of the
Lions and the well-preserved mosaic
floors of the House of Dolphins.
There were ruins related to occupation by the Romans as well as by the
Greeks. The island had no human
inhabitants.

On our first day ashore in Turkey
we landed at Dikili, a small port city

Altar for Worship of Aesculapias
Pergamum, Turkey

with access to Pergamum. This was a cultural center for
the Greeks, and later for the Romans. At one time it is
said to have had a library with two hundred thousand
books, and it was here that parchment for books was
invented. We were told that all of the books were sent to
Alexandria as a gift from Anthony to Cleopatra.

The most interesting place was related to the god
Aesculapias–the one with the snakes. A spring which had
"healing water" had been discovered here, and a hospital
had been built for people who came to drink the water.
There was an altar, with a carving of the symbol of the
two snakes, for the worship of Aesculapias. A group of fol-
lowers of the god, called Aesculapiads, took care of the
patients, stressing exercise, massage and a positive atti-
tude. Over the entrance to the hospital this message (in
Greek) was carved: "IN THE NAME OF THE GREAT GOD
AESCULAPIAS DEATH MAY NOT ENTER HERE." A
tunnel was built for the patients to run through for exercise, and Aesculapiads were said to be in hidden places

along the way chanting, "You're getting better, you're getting better," or some such encouraging phrase. I believe that we chose not to drink any of the water, but we did walk (not run) through the tunnel.

Galen, the famous doctor whose writings about the human body were the textbooks of physicians for several hundred years, was born in Pergamum about A.D. 160. He lived in Rome for a while, then went back to his hometown to practice his profession.

Istanbul was our next stop. We had a drive around the city, then visited some special places. Topkapi, originally called the Seraglio, was the palace of the sultans, but is now a museum. On display is a marvelous collection of jewels—diamonds, rubies, emeralds, and so on. The Topkapi Dagger, set with an enormous emerald, is probably the most famous exhibit. Room after room was filled with shelves of articles of beautiful porcelain. We were told that Suleiman the Magnificent (sultan from 1520 to 1566) used some of the precious porcelain regularly to serve meals to his big family of many wives and children. There were bowls and platters of enormous size.

Before entering the mosques we removed our shoes and replaced them with the floppy cloth ones provided. In the Suleymaniye Mosque our attention was called to the several layers of huge rugs on the floor. When one rug was worn out, it was not removed but a new one was laid over it. We marveled at the magnificent dome and minarets of Saint Sophia from the outside; it is now a museum and we did not have time to see it properly. The Blue Mosque, however, had to be seen inside. The walls, ceilings and columns were decorated with blue tiles of various patterns and shapes. It was amazingly beautiful.

We bypassed most of the crowded market, and went to a rug shop. We were given a description of the way a genuine Turkish rug is made—chiefly in the home of the worker. Designs and materials are supplied, and each strand is individually tied by hand. Length of time required to make a rug is often measured in years. The price is determined largely by the closeness of the knots. I don't remember how many thousand knots per square inch made up one small rug which was made of silk, but the price was many thousands of dollars. It was not a rug which you put on the floor and walk on. A popular size and design was called a prayer rug, used by Moslems to make their prayers facing Mecca. Phebe bought a beautiful one of these to give to her granddaughter for a wedding present.

Each of us succumbed to the temptation to buy a harem ring. Any number from one to five rings indicated status in the harem. One ring indicates the lowest members and five rings fastened together are worn by the favorite wife. I decided upon a four-ring set, but Phebe insisted upon five.

Our guide refreshed our memories about one phase of Turkish history. Kemal Ataturk, in the nineteen-tens and early twenties, successfully led a movement to abolish the sultanate and establish a republic. As president of this republic, he set upon a course aimed at westernizing Turkey. Among the changes he made were abolition of polygamy and the wearing of the veil by women and the fez by men.

She then gave us a tongue-in-cheek picture of current life in the country. After having pointed out several coffee houses in which men were congregated, she explained that they have to get together every morning to settle the problems of the country, and in the afternoon to do likewise for the problems of the world. She feels sorry for them because they have this burdensome task, and also because they have never adjusted to having only one wife. Meanwhile, their wives have a very happy and easy life. All they have to do is look after the children and the house, cook the meals, plow the fields, and do whatever job they can find to earn money for food and clothes for the family. She concluded by saying that she is "happily divorced."

Our next stop was Kusadasi, a resort town from which we went to Ephesus. This city was founded in 1100 B.C. by Athenians and became an important trade and cultural center. One of the famous buildings, the temple of Artemis (Diana, after the Romans conquered the area), is named as one of the wonders of the ancient world. St. Paul and St. John preached there, and St. John was buried there. As a result of destruction caused by a series of conflicts and then the silting in of the harbor, the city was left in ruins for many years. Excavations and restorations have taken place, and evidence of a number of buildings may be seen. One noticeable one was evidently a library. The enormous amphitheater, built by the Romans and said to have been the site of some of the encounters between Christians and lions, was very well preserved.

Our landing in Israel was at Ashdod, where we took a bus for Jerusalem and Bethlehem. On the way, our Jewish guide insisted on stopping at a place called "The Elvis Presley" where Elvis memorabilia was offered for sale. We did see the Garden of Gethsemane and the Mount of Olives, with many olive trees looking as old as the two thousand years attributed to them. The Church of the Holy Sepulcher was built over some of the sites of the Via

Dolorosa, and it was confusing to try to relate the various points to the concept I had of the events which took place there. It was also disappointing to see the church poorly kept and dirty. The Church of the Nativity in Bethlehem was kept in better condition, but the little cave which was labeled the birthplace of Jesus was not in keeping with my concept of the event.

In Jerusalem we saw the Wailing Wall, sacred to the Jews, with crowds of women at their section of the wall and a scant number of men at their area. We did not, of course, go into the Moslem section of the city, but the mosque called Dome of the Rock towered prominently over everything.

Our ship next docked at Port Said, the northern terminal of the Suez Canal. From there we went by bus to Alexandria, founded in 332 B.C. by Alexander the Great. It became a center of culture and learning, with a famous library, the holdings of which were enhanced by Anthony's gift to Cleopatra. The city has been the victim of many battles and changes of ownership, but is now a beautiful and busy seaport. We visited a museum with artifacts representing the city's varied history. The site where once stood the Pharos lighthouse, another wonder of the ancient world, was pointed out to us.

We had a stop at Cairo, where we went to the top of the citadel built by Sultan Saladin, and had a brief tour of the Cairo Museum. Then, of course, there was a visit to the Great Pyramids and the Sphinx. (I must say here that this was my second visit to Cairo, but because this trip started in Europe I have mentioned it here. More details of my first trip to Egypt will be included in the section on Africa.)

Our time spent on the cruise ship during this tour was especially pleasant and restful. We were assigned a table with two very congenial couples, and it seemed that other passengers and the crew members were unusually friendly. The food was delicious, and we were honored one evening by an invitation to have dinner at the Captain's table.

Switzerland by Train, 1990

Georgia's nephew, George Drury, is an expert on almost anything regarding trains. As a sideline to his other work he writes articles about trains and serves now and then as study leader for Smithsonian tours by train. Georgia and I participated in one of these tours which took us on some interesting adventures in Switzerland in September, 1990.

We had a flight to Zurich and then a bus ride to Lucerne. A tour of the city ended at the Transportation Museum, with examples of a variety of old and new types of trains on display. We then had an interesting type of movie in which we were seated as if in a train. We had the sensation of riding up and down mountains, passing other trains, switching tracks, an emergency stop, and so on. After that we had a real train ride to the top of Rigi, a scenic mountain nearby.

We boarded a train from Lucerne to Brienz, with its beautiful lake. We changed there to a cog railway with a steam-driven train which pushed us from the rear to the top of Mount Rothorn. Beautiful trees and many colorful flowers lined the sides of the mountain, but the top was shrouded in heavy fog. A diesel-powered train brought us down. This was representative of many changes of trains during the entire trip. Different terrains require different types of tracks, gauges, and engines.

A lovely boat ride took us across Lake Brienz to Interlaken. There were several settlements along the way with chalet-type houses and window boxes of flowers. In our Interlaken Hotel, George presented a slide lecture explaining the different types of trains and tracks.

We changed trains twice before getting to the cog railway which took us to the Jungfraujock. This is a sort of glacial saddle near the Jungfrau, and the station there is the highest one in Europe.

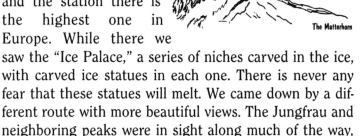

The Matterhorn

While there we saw the "Ice Palace," a series of niches carved in the ice, with carved ice statues in each one. There is never any fear that these statues will melt. We came down by a different route with more beautiful views. The Jungfrau and neighboring peaks were in sight along much of the way, as well as from the city of Interlaken.

On the way to Zermatt we saw more beautiful scenery, as well as the inside of several tunnels. Cars or buses are not allowed in Zermatt, and it is very peaceful and inviting. We had a beautiful room with a view of the Matterhorn. After a restful night we set out (by train, of course) to get a closer look at this peak. This, alone, would have made the trip worthwhile. It was magnificent!

In the evening, George gave a lecture entitled, "They Say It Is the World's Slowest Express," to prepare us for our trip the next day on the *Glacier Express*. It took us along the Rhone River for awhile, and there was a spectacular gorge on the way. There were other beautiful valleys and snow-covered peaks, as well as several tunnels. In one exceptionally long tunnel the track spiraled upward and came out at a higher level than it entered. Since this was a day-long trip, we walked through many cars to get to the dining car for lunch. Our destination was St. Moritz.

In St. Moritz we visited the train dispatch and service station. Service and repairs are made here, and locomotives are assigned to their routes. A control center keeps track of where every train is at all times, and transmits any warning messages, as well as routine information about meeting another train and where and when to take a siding. We saw snow-removal and other service devices. Later on we took a ride to the top of Muotas Murangl, a mountain overlooking St. Moritz and its three lovely lakes.

The next day we took the *Bernini Express* to Tirano, Italy. On the way the train made an unusual loop. We passed over a viaduct then made a wide circle and came back, this time going under it. In Tirano we took a bus for Lugano, riding on a rough and crowded Italian road for a while. At one point we went along Lake Como and stopped at the place where Mussolini was said to have been apprehended when he tried to escape after his downfall in World War II. He and his mistress were shot near the spot at which we stopped.

Back in Switzerland the roads were better and we reached the beautiful city of Lugano, on Lake Lugano. We enjoyed walking by lovely gardens and visiting an art gallery. In the evening George's lecture was entitled "Which *Oriental Express* Do You Mean?" He told us something of several great trains of Europe and prepared us for our ride in a luxurious *Orient Express* car to Zurich the next day. From there we departed for home.

Greenland, Iceland, Orkneys, 1992

This trip in August of 1992 was another one organized by the Smithsonian Institution. It was basically a cruise, with stops along the way. I couldn't find anyone who wanted to go with me so I went as a single. The group met in Boston and flew to Sondre Stromfjord in Greenland. One of the men in the group told of an interesting coincidence. When he read about the tour and noticed that the plane was to land at this place on 9 August he realized that, exactly fifty years before, he had landed a Marine Corps plane on that very spot. This gave him an incentive to go on the trip. He was, of course, on duty in World War II when that happened.

Our first glimpse of Greenland was that it was made up of barren hills, some high enough to have snow on the top. There were some small plants and grasses in the valley but no trees. A bus took us to the port at which our ship was waiting.

The ship had a capacity of one hundred forty passengers and had ninety officers and crew. We started early to hear the various experts who were to be our lecturers on the trip. During the next eight days we had lectures on the Inuits, the history and archeology of Greenland and Iceland, geology, birds, plants, the weather, ecology and so on. Some of them were interesting, but I did not take any notes and would not attempt to quote anything that was said. What we actually saw and heard at the places we visited was more interesting and easier to remember.

For our first trip ashore the ship docked at Nuuk, the capital of Greenland, with a population of about fifteen thousand. The most noticeable thing was that the buildings were all painted in bright colors—red, green, blue or yellow. We went into a church which had a small ship model hanging from the ceiling, symbol of their dependence upon fishing. At an arts and crafts shop we watched women making parts of the national ceremonial dress. Boots and pants were made of sealskin, decorated with lace and lavishly embroidered with beads. The rest of the outfit was of white and flowered fabric, with lots of red braid and more lace and beads. We were told that one must order this costume a year ahead, since it takes that long to make one. A woman and child were there dressed in the finished product and we could, for a price, take a picture of them.

The University of Greenland has fifty students and four faculty members. They teach language, literature, economics and history, but science equipment is too expensive. Students having higher ambitions of study may go to Denmark. The buildings were originally part of a Moravian mission, which was begun in the 1700s.

A buffet dinner at a Nuuk hotel included lots of salads, pâtés, and raw fish; meats were labeled, and included whale, reindeer, seal, veal, salmon and chicken. The whale tasted all right, but was a little tough. Seal was better. Back on the ship we were entertained by a man who played some sort of skin drum, and sang doleful songs.

The next day we were introduced to "zodiacs," which took us to the fishing village of Qequertarsuatsiaat. (Honestly, that is spelled correctly, but don't ask me to pronounce it.) A zodiac is a rubber (or possibly plastic) boat which can be inflated to look like a large oval-shaped doughnut with a bottom. It was tricky to get in and out of, but a little practice helped. At this village we saw a fish-processing place, a retirement home where the residents were selling craft ware at unreasonable prices, and more colorful houses. One of our botanists on the ship selected a "flower of the day" at each trip ashore, and that day it was a dogwood. Would you believe a dogwood that was only six inches high? It had the right kind of flower, but the short growing season makes it do its thing in a hurry.

The only other place we went ashore in Greenland was Arsuk. On the way, our zodiac driver took us as near to an iceberg as it is safe to go. It was beautiful. Arsuk is a very small fishing village, but we saw there the kind of winter home the Inuits sometimes use. It is a long tunnel with an arched top covered with sod. It reminded me of the ride we used to see at fairgrounds called a caterpillar. I don't know how it is ventilated, but I think it would be stuffy. On this excursion it was very cold and windy, and after we got back on the ship rough weather set in and stayed for three days. We missed the last scheduled stop on Greenland, and the time was taken up because an extra day was required to get to Iceland.

Perhaps it should be mentioned that for a long time Greenland and Iceland were under the control of Denmark. Iceland became an independent republic in 1944, and Greenland has had a position comparable to a state of Denmark since 1979.

Iceland is best known for its geysers and boiling hot springs. It makes use of underground hot water to warm its homes, and to provide some of its energy. There are also many waterfalls which are sources of energy. In Reykjavik we saw greenhouses which are heated by underground water and produce vegetables all year. We also saw a zoo, and the house in which Reagan and Gorbachev held their much-publicized meeting.

We rode in a small boat around the Snaefellnes Peninsula with its interesting caves and cliffs. Many kinds of birds, including large numbers of puffins, were nesting on the cliffs. In the geyser area we saw many boiling springs, puffs of steam, and a geyser which spurted up a column of steam every eight to ten minutes.

One very interesting sight was the rift valley, where continents are very slowly drifting apart, making a deep gorge, and increasing the size of Iceland.

At Westman Island we visited the town of Heimeay and heard its story. In the middle of the night of 23 January 1973, the mountain above the town started sending down fire and molten lava. The inhabitants (about thirty-five hundred) were warned, and all escaped to the main island by fishing boats and help from the United States Air Force. The lava covered the town and threatened to fill the harbor. The United States Navy used planes to spray thousands of gallons of water on the lava and succeeded in cooling it enough to save the harbor. After several months many of the inhabitants came back and built new houses on top of the ruins. All of the people, including children, worked to clear the dust from the surrounding hills so that vegetation could be replaced. Previously the town had depended on rainwater, but after the eruption this was poisonous, so a new system was set up. Glacier water was piped from the mainland, and some of this was then run through the heated underground to be warmed, so that now the people have heat for their homes, as well as safe water.

One amusing story which came out during the excitement of the evacuation concerned the people in a rest home for old people. A pilot came to get them and rushed them into a plane. When he got them settled in safely, they complained that they couldn't eat because they had left their teeth, which they had taken out for the night, back in the home. In the nick of time, the pilot went back and collected all the teeth in a bag. Then there had to be a session of "trying on" teeth to get a set that would fit.

Another reminder of the changes in the earth was a view of Sturtsey. This is an island formed by underwater volcanic eruption which continued to build up from 1963 to 1967. Our ship took us all around it, and it is quite a large island. So far, it appeared to be completely barren.

Cruise ships are noted for an abundance of food. They start with an eye-opener, then comes breakfast, mid-morning snack, lunch, afternoon tea, dinner, and a midnight snack. I try to limit myself to three regular meals, but I want to list the items on the menu of one dinner—the Captain's farewell dinner: escargot, seafood soup, fish course of salmon, rice and vegetables, raspberry sorbet, salad, main course of beef Wellington, flaming baked Alaska, and an assortment of nuts and candies. Of course there were crackers and bread, beverages, and various kinds of garnishes. I was glad I didn't eat much of the first courses, because the Beef Wellington just about tops the list of all the best foods I have ever tasted.

The Orkney Islands belong to Scotland and are located near its northern coast. The capital, Kirkwall, is noted chiefly for its church which was built by the Vikings in the twelfth century. Near the town there are interesting remains of a village in which people are thought to have lived four thousand years ago. A heavy storm in 1850 washed away much of the beach, and uncovered evidence of a village of stone houses, complete with stone furniture and other clues to the lifestyle of an ancient era. At some distance away there is a large mound which was obviously a communal burial chamber, and estimated to date back to 3000 B.C. Then there is the Ring of Brodges, evidently at one time a large ring of vertical stones, of which several still stand. Their purpose is not known.

From the Orkneys we went to our ship and on to Aberdeen from where we started on our series of flights home.

❈ Trip with Cape Fear High School Group, 1995

The chorus and band of Cape Fear High School were selected by Governor Jim Hunt as the official representatives of North Carolina in the World Liberation Music Celebration in Europe. The invitation to this portion of the observance of the fiftieth anniversary of the end of World War II was signed by the mayor of Paris, the lord mayor of London, and the Minister of Veterans and War Victims. Fifty-three students, some of whom performed in the chorus and the band, were accompanied by thirty-three adults, including the directors, some parents, and friends. I was pleased to be among those invited to go.

The first stop was a three-day stay in Paris. The groups performed in two city parks, in which placards had been placed to announce who they were and why they were in Paris. They were well received by the gratifyingly large audiences. They also gave a beautiful impromptu rendition of Lutkin's benediction, "The Lord Bless You and Keep You" in magnificent St. Chapelle. There were drives to point out some of the outstanding sights of the city, with stops at several places, including the Louvre, the Arc de Triomphe, and so on. Then there were tours of Versailles, and chateaus of Blois and Chambord in the Loire Valley.

The most memorable occasion in France was a service at the American Memorial complex in the Omaha and Utah beach areas of Normandy. A colonnade faces a cemetery with the graves of nine hundred Americans who lost their lives there. The service began with a recording of our national anthem, followed by a twenty-one gun salute and the playing of "Taps." A member of our group who was formerly in military service then placed a wreath at the foot of an impressive statue of a young man. Facing the cemetery, our band played a medley of the theme songs of each of our military units, and the chorus then sang the "Armed Forces Hymn" and the "Benediction." After the service we went to Caen to visit the Peace Memorial. There we saw movies showing preparations for the D-Day Invasion as well as some stark photography of the actual event. The happenings of the entire day were very moving and unforgettable.

Next on the schedule was Brussels. There were some very good tours of the city, with stops to see a lace factory, partake of some genuine Belgian waffles, and to walk around the city square with its impressive historic buildings. An outing to Blankenberge Beach, on the North Sea, preceded the students' performance at the community center in the town. A convention of senior citizens which was in progress provided an appreciative audience.

Amsterdam was next on the agenda. Our guide on the city tour gave interesting information about the history and lifestyle of the people, especially as it related to the development and influence of the canal system. Stops were made for visiting the Van Gogh Museum, the Anne Frank House, a wooden shoe factory and cheese factory. We also had a pleasant cruise on some of the canals.

From the Netherlands we boarded a ferry boat at Hoek Van Holland for a trip across the North Sea to Harwich, England. This ferry was quite a contrast to those we have from Hatteras to Ocracoke. It was a very large ship, with seven decks in addition to the lower ones which were carrying a number of large trucks, buses and cars. All sorts of conveniences were available, including a movie theater, shops, lounge areas and restaurants. The trip required nine hours, and even though we had been told that the North Sea can be very rough, our voyage was calm and pleasant.

Buses from Harwich took us to London, where our first stop was for fish and chips at Flanagan's on Baker Street. The next day we went to Windsor for a tour of the beautiful and famous Windsor Castle. From there we went to Corsham, where the students were to perform in the evening. An art festival was in progress in the town, and some local bands were also on the program. Our chorus and band made a very good showing.

The most interesting feature of our stay in Corsham was that our entire group was assigned, in twos, threes or fours, to host families for the night. Four of us adults were invited to the home of the director of the local festival. Parts of her home had been built in 1719, and it was located in the historical preservation district. There were beautiful furnishings, some of which were interesting antiques. We had tea in the attractive garden, and later had dinner there. The dinner consisted of four pies and several salad items. There was a beef pie, a pork pie, a turkey and mushroom pie, and a cheese and vegetable pie. All were delicious. After the performance, we had dessert with the husband, who is employed with the defense department.

Before leaving for London the next day we had an opportunity to see some almshouses and a school for poor children built in 1648 and still in good condition as museums. The large and imposing teacher's desk in the school room had a wooden hand carved on each side. We were told that the one on the right was for a candle, and the other held a ferrule which was used for punishing children who misbehaved.

Back in London, where the temperature hovered around one hundred degrees, we had a brief tour on a bus with no air-conditioning. The only stop was at Westminster Abbey. We had tickets to a play that evening, but I decided to let the young people enjoy it without me. They reported that the theater had air-conditioning, but the hotel room did not.

The next day we went to Stratford-on-Avon, and not only did the bus have no air-conditioning, it had no shades for protection against the sun's rays. Since I had seen Shakespeare's tomb and Anne Hathaway's cottage, I tried to find a shady place while the group did the tour.

We went directly to Battersea Park when we returned to London, for the last scheduled performance of the trip. I know that the students were hot and tired, but they were in good spirits and gave it their best effort. One of the girls told me that they were inspired because some of their members had graduated, and since this was to be their last appearance with the group everyone wanted it to be special. There were lots of hugs and tears afterward.

I must say something in tribute to Betty Neill Parsons, choral director, and Jim Crayton, band director, for their excellent management during the trip. Of course the students had been well trained in their performance skills or they would not have been selected for this venture, but they were also organized in the mechanical aspects. Several times they had to change their costumes in buses (one bus for boys and one for girls, of course) but I never heard a complaint. Each one knew what his special job was in unpacking and setting up instruments, stands, chairs and music sheets, and it ran like clockwork. They were all courteous and helpful, and the adults charged with checking and overseeing did their job with no obvious problems. It was an uplifting experience for me.

Australia and New Zealand, 1979

This trip in November 1979 was planned for Delta Kappa Gamma members by the International Travel and Study Committee and lasted about five weeks. The first stop was at Fiji, and from there we went to New Zealand, then to Australia, back to New Zealand, and had a few days in Tahiti on the way home. Phebe and I knew several of the thirty-two members already, and all of us became close friends as the days passed.

We landed at the airport in Nadi, and spent four days in Fiji. On the way from the airport we enjoyed seeing beautiful tropical trees—poinciana, acacia, African tulip, frangipani—and many other flowering plants. We were met at the hotel by men wearing colorful skirts and a flower over one ear. Most of the day was spent resting from the flight and enjoying the lovely view of the ocean from our hotel window. Meals were served outdoors, and the breakfast and lunch were great, but the buffet dinner was unbelievable. There were beautiful displays of fruits and salads; meats included curried fish, chicken, fish with coconut, lamb chops, steak broiled to one's taste, and a whole roasted pig; there were lots of cakes and puddings for dessert. Everything was served in fancy dishes, and flowers and fruits were used as decorations. It seemed a shame to disturb all of this by eating some of it, but it tasted as good as it looked.

After dinner we were entertained by a "meke"—singing and dancing by Fijians in native costume. It began and ended with a conch fanfare. A gala evening!

The next morning I had a tangerine at breakfast and fed the seeds to a mynah perched on the table. After breakfast we boarded a bus for Suva, the capital of Fiji. Along the way there were lots of sugar cane fields. We learned that a crop requires eighteen months to mature, and many farmers plant a field every six months so that stages of growth are staggered. At that time sugar accounted for sixty-three percent of the country's income, with tourism ranking second. Most of the people live in villages, and tin roofs were beginning to replace the traditional thatched ones. Each village had a chief, and a church. Education was not required, and some children had to walk a long distance to get to a school in another village.

Other information we received along the way concerned the Indian inhabitants. When England took over the islands in 1870, the Fijians refused to work in their fields. The English proceeded to bring people from India, presumably to work five years and then go back home. However, about two-thirds of the total 63,000 who were brought at intervals decided to stay. The natives and the Indians had lived separately through the years.

Another story concerns a missionary who came over when cannibalism was still in practice. Some of the tribes accepted Christianity and gave up the practice, but one tribe was difficult to persuade. One day the missionary made the mistake of touching the head of the chief of this tribe, and for punishment he was killed—and eaten. Touching anyone's head is still considered an insult, but it doesn't demand so severe a penalty. Children tease one another and have fights over touching one another's head. Believe me, there was no way that any of us would have dared touch the head of one of those people.

145

We stopped for lunch at a resort on the beach. There were large hotels and beautiful grounds. Fiji became a crown colony of England in 1870, and a member of the British Commonwealth in 1970. Prince Charles came to help celebrate that event.

In Suva we walked in fabulous botanical gardens and visited a museum with war canoes and other mementos of the past. Then we were invited to a kava ceremony. A drink is prepared from ground roots of the kava plant, in a procedure carried on by specially qualified persons and using special vessels. Drinking it is a kind of ritual, but I don't remember the details. It is not alcoholic, but is said to cause the tongue to get numb. Being well-mannered people, we did not refuse to taste it, but we did not drink enough to feel any effect.

We watched some of the craftsmen making tapa cloth (from the bark of the pandanus tree), pottery, baskets and wood carvings. We also went to a beautiful orchid garden, with some plants donated by Raymond Burr. (He owned a small island nearby, and was an avid orchid grower.)

Coconut Trees—Fiji

Next we had a trip in a glass-bottom boat, and saw some unusual marine animals and plants. There was a stop at Mosquito Island to observe demonstrations with coconuts. A barefoot young man climbed a tall tree to get coconuts, and we sampled the milk and the fruit. Then some of the young people made hats, baskets, belts and birds from the leaves.

I have never been in a country in which people were as friendly and as happy as in Fiji. They seemed to truly enjoy life and wanted us to join them in having fun. We left there with a special affection for them and the hope that their country would continue to be peaceful and happy.

Our first stop in New Zealand, at Auckland, was for only one day, but we met the knowledgeable and witty guide who was to be with us during all of our New Zealand experiences. His name was Les (pronounced "Liz"). He showed us around the city for a while and then took us to our hotel. During the afternoon we walked across the street to the campus of the University of Auckland, and enjoyed the display of native and introduced trees and other plants.

Our first activity in Sydney was a sightseeing tour of the city. We admired the beautiful purple jacaranda trees and the red-flowered lliwara, which blooms only once in six years. In the Paddington area, an early settlement, there were many houses with wrought iron balconies. The iron, we were told, had been brought over from England as ballast for the ships coming for wheat and wool.

The most outstanding and interesting building in the city is the opera house, finished in 1973 at a cost of $102 million. Its unusual roof was designed to resemble sails of a ship. Inside there were four theaters: cinema, drama, opera, and concert. One evening we had the privilege of attending a concert there.

One afternoon we had a cruise of the harbor on the *Captain Cook II*. Australia was, of course, settled in part by convicts from England. The place at which they land-

Opera House—Sydney, Australia

ed, called "The Rocks," had become a container loading area and a dry dock. Pinch-gut Island, where ornery convicts were kept in solitary on bread and water, was pointed out to us. We went under the Harbor Bridge and had a beautiful view of the city.

On one of the tours we went to the Waterloo Tavern, evidently a popular place with some of the early settlers. We saw the trapdoor in front of the bar, sometimes used to shanghai an over-indulger by dumping him into a cellar below. From there he was taken to a ship, and when he sobered up he was somewhere at sea, an unwilling member of the ship's crew.

We had a short trip out from Sydney, called the "Jolly Swagman Tour," to Dubbo. First we attended a cattle show, and then went to a sheepshearing. The shearing was done with electric shears, and required about three minutes per sheep. The wool came off almost entirely in one piece. Next we went to a tannery where hides were prepared for leather.

Kangaroo

In a very interesting zoo we were allowed to walk around and feed kangaroos and emus. The emus were greedy but the kangaroos seemed bored with it all. There were other animals as well, and many kinds of birds. The keeper brought out a snarling wombat. This animal, about the size and shape of a pig, lives in burrows. Unlike other marsupials, the pouch of the female opens upward, from the rear—a wonderful accommodation to keep soil from getting into the pouch when the animal is burrowing.

Another feature of this tour was a sheep roundup, an interesting demonstration of the work of the trainer and the sheep dogs. The most appealing part of this venture was seeing and even holding some of the baby lambs.

I must mention that on our drives anywhere in Australia we saw many startlingly beautiful trees. Some were called "bottle-brush" because this is what their flowers looked like. They came in red, yellow and pink varieties. Another was called "silky oak" but with its silky yellow flowers it had no relation to our oaks. Then there was the national tree of Australia, the yellow-flowered wattle tree. Of course we often saw representatives of the three hundred varieties of eucalyptus.

A dinner with speeches by the Dubbo celebrities ended our Jolly Swagman Tour.

Our next venture was to Alice Springs, where we stopped only briefly before setting out, in small planes, for Ayers Rock. We had a very hot bus ride around the Rock, but were interested in hearing about its geological history. There were occasional stops at caves to see examples of aboriginal art. We were impressed by the colors of the rock as the sun went down, and got up early in the morning to see the changing hues as the sun rose. Our overnight accommodations were adequate, but far from elegant. The manager apologized for the lack of variety in the food, but explained that their only source of anything was a plane such as the ones we came on. They did, mercifully, have air-conditioning.

Back at Alice Springs we had a tour of the area. Water could not be seen in the spring because sand had washed in; however, water could be pumped out if a pipe was inserted to a depth of six feet. This site had been discovered in 1870 by a scout for a telegraph company, and was selected for establishing a station. The telegraph had limited capacity, and northbound messages were sent one day and southbound ones the next. When it first opened there was a twelve month waiting list for messages to Europe. Capability has been enhanced, and the station was a pioneer in using solar energy. Alice Springs has also become an important railway center.

An interesting visit was made to the Royal Flying Doctor's Station, begun in 1929. Donations had been solicited so that two-way radio stations could be installed at all outposts. A patient's symptoms can be described to the doctor, who decides whether to prescribe medication, go to visit the person, or send one of the flying ambulances to bring him to the hospital. The most distant outpost is four hours away. Each outpost has an airstrip and a medicine chest, all alike. Children in an isolated area may use the two-way radio system if no school is available.

From Alice Springs we went to Melbourne. The greatest attraction there was the nearby colony of fairy penguins. These, the smallest species of penguins, are about one foot high. They nest in burrows in banks near the sea on Phillips Island. During the breeding season—August to

December–the male and the female take turns either sitting on the nest or caring for the chicks, or going to the sea each morning to catch fish. At dusk there is a parade of thousands of them coming from the water and going to their burrows to regurgitate the fish for the mate and chicks. They are unbelievably cute as they waddle directly (usually) to the proper burrow. Sometimes the mates make noises as if to guide them, or possibly to welcome them home. We were warned to be careful not to step on one of them, as they walk on or between your feet if you are in their path. The Phillips Island colony is the largest known one of this species–about five thousand in number.

Another Melbourne attraction was an establishment specializing in opals. The largest known opal, seventeen thousand carats, was on display there. There were demonstrations of different types of opals and how they are cut and made into jewelry. Prices ranged from inexpensive to out-of-this-world.

There was a wildlife preserve in which we saw lots of animals and birds, but the real treat was seeing a platypus.

The island of Tasmania, a state of Australia, was next on our agenda. I remembered it from biology lessons for its association with a number of rare or extinct animals. The only unusual ones we saw in their natural habitat were an echidna, a wallaby and a Tasmanian opossum–very much like ours except that it had a bushy tail. In a zoo, the main attractions were the Tasmanian devil and the wombat, each of

Koala

151

which had young ones. We were allowed to hold the baby wombat and have our picture taken with him. The little devil was so vicious that the zoo keeper had to hold him— carefully.

Port Arthur, now a pleasant little town, was founded as a penal colony. The worst criminals were sent there, sentenced to hard labor or solitary confinement. The keepers of the prison prided themselves on the fact that no one had ever escaped, and there were stories of cruel punishment to anyone who dared to try. In the 1880s Queen Victoria found out about this prison and ordered it to be demolished, as it was a disgrace to England. Some sections still remain, however, along with some mementos of inhuman punishments.

Our guide and bus driver in Tasmania, named Bob, was a talented singer, and he entertained us now and then by singing songs such as "Waltzing Matilda," "The Jolly Swagman," and "Tie Me Kangaroo Down"—all of the verses. He prefaced our favorite number by a story of his experiences in New Guinea following World War II. After the fighting was over he and some other survivors had to wait a good while for transportation home. They had lots of time on their hands and no entertainment except a wind-up record player and a record of "When Day is Done."

Somehow they befriended a native boy, and gave him the job of winding the record player, while they lolled around and listened to that song over and over. Now and then the player would run down and the boy would wind it up without removing the needle. Bob's version imitated this by gradually getting slower and lower in pitch, then speeding up and raising the pitch back to normal. We called for several encores on that one.

Incidentally he expressed great admiration for the American soldiers he had met during the war, and for our country in general. He said at one time, "If it had not been for the United States, I would be speaking Japanese to you now." He sang "Now Is The Hour" for us as he drove us to the airport for our flight back to the mainland.

We returned to Sydney for a flight to Christchurch, New Zealand. One beautiful sight near our hotel was the Avon River with rhododendrons in bloom along its banks. There were many lovely parks and gardens. In one museum a section was devoted to Antarctic explorations, many of which were based in Christchurch. One of our tour arrangements was a dinner in the home of a local family. We enjoyed meeting and talking with these people, and the dinner was delicious. I can remember that we had roast lamb, and *pavlova* decorated with slices of kiwi fruit, among many other good things.

We left Christchurch for a visit to Mount Cook. For a while we rode along flat or rolling land with many sheep. There are eight sheep for every human in New Zealand. Next we went through rough and hilly territory, and finally reached the Southern Alps. There was a stop at a small airfield so that we could have a flight in an eight-passenger Cessna over the Tasman Glacier, elevation six thousand feet. It was a little scary, as the pilot made sure that we saw the glacier from several angles. It was an interesting flight. The scenery was beautiful all the way to snow-capped Mount Cook, highest peak in New Zealand at 12,349 feet.

Another scenic drive took us to Queenstown, where we saw a cattle show. Representatives of each breed in the country paraded, as a speaker told of the characteristics and use of each breed. Our next stop was Te Anau, where we celebrated Thanksgiving. There was a special dinner, including turkey (the alternate meat was rabbit), cranberry sauce, sweet potatoes and pumpkin pie. The chef came out to get our reaction, and was pleased with our response. He said that for two weeks he had been getting suggestions for this meal from American tourists.

On the way to Milford Sound we passed through several changes in vegetation, including a rain forest with three hundred or more inches of rainfall each year. Milford Sound, actually a fjord, is rightfully considered one of the most beautiful places in the world. We had a boat ride to its outlet at Tasman Bay and back, marveling at the spectacular cliffs and waterfalls. The boat passed behind one of the falls for an added bit of excitement.

We went back to Queenstown, where we had a gondola ride to the top of a mountain for more beautiful views and lunch. The next place on our agenda was Rotorua. This area is known especially for its Maori population and its hot springs and geysers. We went to a Maori village and saw an arts and crafts center, a meeting house, and a church with a stained glass window picturing Christ in a Maori robe. There were many geysers, one of which spurts up hot water and steam to a height of as much as 125 feet. The Maoris use the underground heat for cooking, by burying a pot of food in the ground.

At Rotorua we attended a sheep show, similar in routine to the cattle show in Queenstown. The highlights of the animal kingdom, however, were the tuatara and the kiwi. The tuatara, related to an ancient group of reptiles, is the only surviving species with a third eye. It is found only in a limited area of New Zealand. The kiwi is better known by its pictures than by sightings, because it is nocturnal. A trick of reversing day and night makes it possible to see them in a zoo. During the day the enclosure is darkened, and bright lights are

New Zealand Kiwi

turned on at night. After we got used to the dim light we could see them quite well. They, again, are native only to New Zealand. Kiwis eat worms and grubs, which they locate easily because their nostrils are located at the tip of their long bills. They weigh about four pounds, and the female lays an egg weighing half a pound. The male sits on the egg, which requires seventy-five to eighty days to hatch, and the female brings home food during this time.

We had a special Maori dinner offering lamb, venison, wild pig, fish (cooked or raw), and smoked eel. Dessert was an assortment of fruit. The hosts were dressed in native costumes, and we were entertained by native songs and dances. They also told us about their history and customs and the significance of some of their symbols.

On the way to Auckland we had a stop at Waitomo Caverns. There were many interesting formations but the outstanding sight was the Glowworm Grotto. In order to see the myriads of lights of these tiny creatures we had to be in total darkness. It was slippery underfoot, but we managed without accident.

Among the sights in Auckland the most impressive one was from the top of Mount Eden, an extinct volcano almost in the center of the city. It is a viewpoint for the entire city and the coasts on either side. Other places we visited were a greenstone (variety of jade) factory and sales room, and an interesting winery.

❊   ❊   ❊   ❊   ❊

During our first meeting with the group, one of the members showed us a brochure advertising a one-day flight over a part of Antarctica. It was scheduled for the last day we were to be in Auckland, and our friend persuaded three others in the group to go with her. On the night before the flight, they were very excited as the rest of us wished them well. They were to meet us in the Auckland airport for our flight to Tahiti the next night.

Les and our Delta Kappa Gamma guide learned during the afternoon that contact with the Antarctic plane had been lost, but they didn't tell us at that time. Les took us on an unscheduled trip to the top of Mt. Eden so that we could see the city lights and the stars. We understood later why he wanted us to have this beautiful memory of Auckland. When we got to the airport, we found out that the plane to Antarctica was presumed lost. We had to go on to Tahiti, and during the night we got the message that the wreckage of a plane had been sighted and that there was no hope of survivors. When we got to Tahiti the next morning it was difficult to respond properly to the cheerful greeting and the leis presented us by the Tahiti welcoming committee.

Tahiti is a very beautiful island with a wealth of tropical flowers and trees. It had been considered a tropical paradise by many people. Among its inhabitants have been several authors, and the artist, Gauguin. We went through the Gauguin museum, but I can't say that I had much enthusiasm for his paintings. Following an early expedition by Captain Cook, the British claimed the island, and during the United States Civil War they grew cotton there and brought Chinese people in to work in the fields. The French later won the island from the

British, and the population was about forty percent native, ten percent Chinese, and the rest mostly French.

Our hotel was built into the side of a cliff on the ocean, and the floors were numbered from one to ten downward. There was an outside stairway of 275 steps which Phebe and I walked down and had a stroll on the black-sanded beach.

For several reasons—including our tendency to compare these people, to their disadvantage, to those of Fiji, our sadness over the loss of our four friends, and probably the fact that we were simply ready to go home—the stop at Tahiti did not serve as the exhilarating finale to our trip for which the planners had probably hoped.

$\mathcal{A}$frica

✖ *Egypt, 1980*

During our tour in Egypt a remarkable guide named El Sayid Mahain told us about the history of ancient Egypt, the rulers, the way of life, the art and language, something of the many gods which were worshipped, burial rituals, and the preparation for life after death. We were introduced to much of this information during our tour of the Cairo Museum on our first day in the country. This was continued as we went to temples, monuments, tombs and other places of interest along the way. This was very informative and I took lots of notes, but I don't plan to go into any of this except briefly as it relates to some of the places we visited.

After the museum tour we stopped at a shop which makes cartouches to order. These are pendants on which one's name is embossed in hieroglyphics. We also went to a papyrus institute, where we saw paper being made in at least somewhat the same way as it was done centuries ago. There were colorful paintings on papyrus for display and for purchase.

Naturally the biggest event in this part of the country was a trip to the Pyramids of Giza and the Sphinx. The only accepted way to get up the hill where these monuments are located was by camelback. The camels were decorated with colorful harness and tassels, and the drivers wore the customary djellabahs and turbans. Georgia and I had planned to take pictures of one another, but

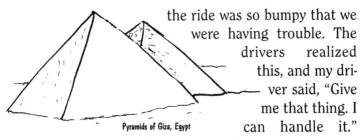

the ride was so bumpy that we were having trouble. The drivers realized this, and my driver said, "Give me that thing. I can handle it."

Pyramids of Giza, Egypt

Sure enough, I think that picture is the best one I got on the trip.

The pyramids looked like their pictures, except that they were much larger than expected. The base of the Great Pyramid of Cheops, largest of the three at the site, covers thirteen acres and is 450 feet high. The Sphinx was undergoing repairs and had a lot of scaffolding around it. Through the years the neck had been weakened, and a United Nations commission was making an attempt to strengthen it.

In a very poor section of Cairo, dirty and dilapidated, we saw an old Coptic Christian church. This church is said to have been built over the cave in which the Holy Family lived during their exile to Egypt to escape the decree of Herod. We went down into the basement area which we were told was originally the cave. The entire building and the furnishings were in need of repair and cleaning. There were stories of miraculous events which had helped the people who sheltered the Family overnight as they traveled to what is now Cairo.

On the way to the airport for a flight to Aswan we passed the "City of the Dead." There were thousands of tombs, the more expensive ones with mosque-like domes and others with simple rounded covers. Because of the belief that the "double," or soul of the dead, can return to the tomb, everything must be kept clean and orderly, and some people spend their lives caring for the tombs.

They live in the place and sometimes sleep in the tombs. We were told that there is no way to count these people, but that they are estimated at twenty-five thousand. My notes do not tell me where they get the necessities of life, such as food.

At the airport we had to open our suitcases and empty our purses and personally go through two metal detectors. After they were inspected our bags were sealed. This actually was the practice every time we took a plane in Egypt. El Sayid explained that this extra precaution was taken because Egypt had, only a few days previously, received a diplomatic representative from Israel after several years of severed relations, and that the atmosphere was very tense.

From the plane we could see lots of sand and some dry stream beds, but no sign of vegetation or of human life. Finally we had a view of the tremendous Aswan Dam and Lake Nasser. The dam was built partly to control floods of the Nile, and partly as a source of power for the country. Only six dynamos were in place, although there was provision for many more. A different aspect of the dam intrigued me. El Sayid said that due to the evaporation of water from Lake Nasser, rainfall in the area had increased from practically nothing to a measurable amount, and that adobe houses which had stood for hundreds of years had been destroyed by rain.

We had an exciting ride across the Nile in a felucca (a small sailboat). We couldn't decide whether our crew members were incompetent, or very competent and daring, but while the other boats went smoothly we had a wild ride.

From Aswan there was a short flight to Abu Simbel. This temple, located near the Nile, was much in the news

when it was learned that the dam at Aswan would probably raise the water level near enough to damage it.
Several countries, including the United States, decided to
move it to a location farther up the river. This involved
sawing the walls, columns and statues into pieces, num-
bering them carefully, moving them to the new setting
and putting it all back together again. The temple was
built about 1200 B.C. Its outstanding features are four
seated figures of Ramses II, each sixty-seven feet high and
weighing twelve hundred tons. The lips of each figure
measure three feet across. Inside the temple and facing
the door are four statues of gods, the two center ones
being representations of the gods of the two areas of the
kingdom. The temple was originally oriented so that
exactly on 20 February and 20 October the rays of the ris-
ing sun fell directly on these two sun gods. In its new
location, this phenomenon is still in effect.

Just to demonstrate that people are alike all over the
world, we were told to notice that at the top of the walls
there were some writings carved. This happened, we
understand, during a long period when the temple was
almost entirely covered by sand. The writings were
described as graffiti.

Back in Aswan I went with Georgia to a Saturday afternoon mass at a little Catholic church. It was operated by
French priests and nuns. A priest saw us and spoke with
us, and then did part of the service in English for our
benefit. The church had some beautiful furnishings. We
took a cab to the service, but decided to have a "caleche"
ride back to the hotel. This is a horse-drawn carriage, and
our driver was a toothless old fellow who was very cordial. He wanted to practice his English on us, and he did

quite well with several words. As we passed a mosque, he said "my church."

Later that evening our group went to the local cultural center to see some Nubian dancers. The costumes were beautiful, and the program was quite interesting and entertaining.

The next day we flew to Luxor, farther down the Nile toward Cairo. Our first visit there was to the Temple of Karnak. This is really a number of temples, built at intervals during several centuries from 1200 B.C. The entrance to the temple of Amun-Re was along a lane lined by statues of a type of sphinx. Each one had the body of a lion and the head of a ram. They represented the merger of the two kingdoms of Egypt. Amun, sun god of the northern kingdom, had chosen the lion as his sacred animal, and the ram was sacred to Re, sun god of the southern kingdom. The two gods merged when the two kingdoms united. That is the story, at least. There was a giant statue of Amun-Re in the temple.

Much of the temple had been weathered away during the centuries, but there were columns, walls and statues almost intact. One obelisk was given special attention. It was eighty-two feet high and was covered by carvings. We were told that after such a column had been erected, sand was piled around it almost to the top. Then the carver began work at the top, and the sand was removed as he progressed to the bottom. This particular obelisk originally had a mate which was sent to Louis Napoleon during his rule of France; it still stands in the Place de la Concorde in Paris. In exchange, Louis Napoleon sent a large clock to Cairo to be used in the Great Mosque there, but we understand that the clock never worked.

A great many carvings on walls in the temple represented the life of the ancient people, and most of them
showed the figures wearing clothing. In many cases, however, the laboring men were depicted in the nude.
Something so obvious that it had to be noticed was that
occasionally the male organ of one of these figures had
been chiseled out. No, this was not done for modesty.
Somehow, some people got the idea that this piece of
stone, ground into powder, would have miraculous effects
as a cure for certain illnesses and as an aphrodisiac. This
same kind of damage to such carvings had occurred in a
number of other places we visited in Luxor and in Cairo.

The next day we went on a ferry across the Nile to the
Valley of the Kings. Our first visit was to the tomb of
Amenophis—I believe he was number III. There was a long
tunnel-like passage, lined with carvings and paintings,
and leading finally to the burial chambers. We wondered
how people several centuries ago could have enough light
down under the ground to do all this artwork, since
torches or candles would use the oxygen. The answer was
that they had a series of mirrors, angled in such a way
that sunlight was reflected from one level to another. It
sounds unbelievable, and it would certainly follow that
these productions must have required years to complete.
We learned, however, that each sovereign began to plan
the preparation of his tomb early in his reign. Next we
went into the tomb of Ramses VI, which had even more
elaborate carved and painted figures and larger chambers.

The outstanding attraction was the tomb of King
Tutankhamen. This tomb was actually underneath the
one of Ramses VI, which probably accounts for the fact
that it is the only one of the known ancient tombs which

had not been robbed long ago of all of its valuable articles. Actually, since King Tut was very young and lived for only a short reign, he did not achieve very much. Therefore his tomb is not considered as large or as full of valuable objects as others who had longer and more productive reigns. However, having seen in the museum in Cairo the great display of golden furniture and other articles, elaborate jewelry and enough other symbols of luxury to fill several rooms, it is hard to imagine a greater collection. We saw the chambers into which all of the wealth had been stored. Then we saw the actual crypt from which the body, with its golden mask, had not been removed. This was quite an exciting moment for one who remembered the publicity given the discovery of this tomb in 1922.

We went into the funerary chapels of several of the kings. Evidently the mourning and the burial rites took place in these buildings. Here again there were carvings and paintings. I remember one mural which showed a number of women obviously in mourning. The chapel of Hapshepsut, the only female pharaoh, was being reconstructed.

In the area there were a number of crude shed-like structures where we were told that the current generation of would-be grave robbers live and spend their time digging in the earth under their houses.

Back in Cairo we had a brief visit to Memphis to see some burial chambers which preceded the underground tombs, as well as some pyramids of a type even older than the ones in Giza. They were constructed of several layers, giving them the appearance of having steps. The next day we left Egypt to continue our tour, which was to take us to India.

🌸 *Kenya, 1981*

We were on Sabena Airlines for this trip, and the schedule called for us to go from New York to Brussels, spend the day in Brussels, and then have a night flight to Nairobi. Georgia and I decided that rather than spend a day sleeping we would take advantage of the opportunity to see a little of Brussels. We had some money changed, and after getting concise directions, especially for the return trip, took a bus to the city.

On the way we were saying something which indicated that we did not know anything about the city, and a gentleman from Delaware, in Brussels on business, overheard us. He said he was on the way to see an opera, but was a little early and would be glad to show us around. He thought that we should see the cathedral, a beautiful Gothic one, and then he led us to perhaps the chief tourist attraction in Brussels, the "Mannikin pis." (If you have never heard of this, I'll be glad to tell you about it.)

By then we felt able to find our way by ourselves, so we started at the city square. Since it was Sunday we did not expect to see much except from the outside, but we did find a lace and linen store open. After that we headed back to the bus stop and the hotel.

In Nairobi we spent a day resting and doing a sightseeing tour of the city. There was a museum of wildlife and an interesting archeological display which included some of the Leakey finds. The tomb of Jomo Kenyatta, leader in freeing the country from British rule and establishing a stable multiracial government, was surrounded by four eternal flames.

The next day we had an early morning walk to the city market. There was a great variety of beautifully arranged fruits, vegetables and flowers, interspersed with woven baskets and wood carvings. Later on we organized in groups of six which would continue through the two weeks of our safari by minibus. At one point along the way we had met and spoken to two couples from Ohio, and they asked Georgia and me to join them.

We felt fortunate all through the trip, and were also lucky to have a native Kenyan named George as our guide and driver. He was thoroughly familiar with the areas we were to visit, and we soon learned that he could spot animals at long distances, and would drive across almost any terrain to get near enough for us to see them. Much of the territory was flat plains, but something often passed for roads which we would have called gullies. George had attended a school operated by Seventh Day Adventist missionaries, and his English, though it had an interesting accent, was understandable.

Our first venture was to Mt. Kenya National Park. On most of the way we were passing through bush country with sparse acacia and eucalyptus trees. Then we came to some cultivated fields of sisal, pineapples and coffee. One unusual field had bananas and corn growing together. George was explaining all of this to us, and he informed us that, "The short one are the corn." In this area there were occasional clusters of thatched-roof cottages which were homes of family groups. As we got nearer the mountains, the weather was cooler and vegetation more abundant.

At the mountain lodge we settled in our rooms and then went to sit on the balcony overlooking a water hole to wait for the animals to show up. Actually, a good many different species of animals and birds appeared, but if I named every one we saw at each spot there would be a lot of duplication. I decided to mention only one or two at each viewing and then, at the end of the story, give a complete list of all animals and birds I recorded.

The animal of special interest at this location was the elephant. We had waited for some time and were getting impatient when an elephant appeared through the trees. He stood there looking around for a while and then went back. Then he came out again, followed by a group of about a dozen others, so we decided that he was the scout. We watched them as they drank and bathed, and several other kinds of animals came; a lot of birds visited a feeding station there.

In the morning we checked out the animals at the water hole and had a great view of snow-topped Mt. Kenya. Then we set out for lunch at Lake Nakuru. On the way there were herds of domestic cattle, and several of them decided to cross the road when they saw us coming. There were also wheat fields and other signs of local activity. We had a brief stop to see Thomson's Falls—two hundred feet high and very beautiful. There were a good many native animals along the way, but the big show was the birds at Lake Nakuru. We saw thousands of flamingos and a large group of marabou storks, pelicans and cormorants.

From there we had a ride on a very bumpy unpaved road to the Safariland Hotel near Lake Naivasha. The grounds were beautiful and well cared for. There was a notably large tree called simply a thorn tree. Overnight accommodations were most pleasant.

The road to our next stop, Kishwa Timbo (meaning "head of elephant"), was very interesting. There were beautiful mountains and plains and a volcano which had erupted seventy-five years previously. For a while we were in the Rift Valley and could see the edge of the rift. Later on we passed a satellite-tracking station and communication center. About that time George told us we were entering Masai country.

The Masai tribe, originally a nomadic people, has essentially settled down as cattle herders and farmers on a reserve. The men are typically very tall, with narrow, elongated heads. They wear a sort of tunic and are usually seen carrying a long spear. The story goes that they must kill a lion before being considered an adult, and that they get their strength by drinking a mixture of milk, blood and urine. The women wear a necklace of many strands of beads, long earrings, and a skirt. The small children did not bother with clothes.

As we got farther away from the mountains the vegetation became thinner, with mostly acacia trees and dry grass. This, according to George, is giraffe country, and we saw many of them as well as other animals. The giraffes eat the tops of the acacia trees, and they are sometimes flat across the top. The road was rocky and bumpy, and we were happy when George left it to drive across the plains. Of the animals, wildebeests, or gnus, were most abundant. An interesting secretary bird was spotted.

Our accommodations at Kishwa Timbo were quite interesting. We had large tents, with a small wooden bathroom attached at the rear. All the openings of the tent had super-sized zippers, which we were told to be sure to fasten at night, because the path along the tents was a night trail for animals. We didn't hear any passersby, but baboons played on the top of our tent during the night. We spent two days and nights there and went on several excursions during the time.

On one of our trips in the area we saw two lions, apparently sleeping off a hearty meal they had enjoyed during the night. Our presence did not disturb them, and we were able to drive up about ten feet from them. Then we saw a lioness who most surely had made the kill during the night, and three cubs were playing around her. Incidentally, the minibus drivers knew that if no other animals were near, it probably indicated that there were lions. When one of them spotted a lion, he had a light which he could flash as indication to the other drivers.

At a lake on one occasion we saw several hippopotamuses. There were also white ibis in the area. We passed lots of anthills, some of which appeared to be about ten feet high.

Wart Hog—Kenya

Many of the animals had young ones among them, and wart hogs were the most appealing example. They seemed always to be trotting along, in a straight line, with the father in front, then the mother, and the babies following—all with their tails pointed straight up. We had to say that they looked cute, but when we got a close look at them we realized that they are very homely.

A group of beautiful impalas put on a show of graceful leaping for us as we approached them, and a view of giraffes drinking was entertaining. They spread their front legs as far apart as possible, and lean over to get the water. Then they have to raise their heads in order to swallow. Apparently their throat muscles are not strong enough to make the water go "uphill."

We went back to Nairobi, starting out on another part of the Rift Valley, with a beautiful view of the eastern escarpment. Later on we stopped at a Masai camp for training young men in tribal ways. They go there at about age fourteen, and are allowed to have women family members visit them. Living quarters are the Masai type of house, with rounded roof covered by dung. We saw a number of women and bought some earrings from them. The young men entertained us by doing tribal dancing. The show was not as remarkable as the dress and make-up of the participants. They had attached sisal or wool strands to their hair to make it longer, and it was fashioned into tiny braids. Their heads and bodies were covered by some sort of red paste. Fancy designs on their legs made them look as if they were wearing black lace stockings. They wore short pants.

Along the way George told us that he was thirty-one years old and had worked with tourists for twelve years. He had a wife and three children about a hundred miles from Nairobi, and he lived with his family during the rainy season. While there, he dressed in tribal costume and participated in tribal customs.

We were happy to get to Nairobi and catch up on our laundry and get things straightened out in general. At one time we peeped into the bar and, lo and behold, a group of Masai warriors, wearing their native dress and with swords leaning against the wall, were sitting in there drinking beer.

The next place was Amboseli Park. We did not see many animals along the way because the grass was quite dry, but we did notice a lot of acacia trees with weaver bird nests in them, looking like long-handled gourds. (George's pronunciation: "wivver buds.") We did see, later on in a shady spot, a lioness and cubs feeding on an animal which George identified, by a foot which projected out of the mouth of the lioness, as a young gnu. There were actually several other lions in the area, and Georgia got a beautiful picture of one, with Kilimanjaro in the background. We had several nice views of Kilimanjaro, which is actually in Tanzania. We were very near the border, which was closed at the time.

Acacia Tree with Weaver Bird Nests
Kenya

At a lunch stop we saw a rock hyrax on a wall. This animal is about the size of a large rabbit, but is classified in the same group as elephants because of its five hoofed toes. A sign at the hotel: "Animals are required to be quiet while people are drinking, and vice versa." Later on along the roadside there was one which read: "Watch out for elephants. They have the right of way." We also ran across a large lava flow, from a volcanic eruption about a hundred years ago.

In Amboseli we stayed at Salt Lick Lodge, with buildings shaped like tribal huts and mounted on stilts. The water holes came up almost under the huts, making it easy to see the animals. A group of about fifty elephants came, but the real invasion was by a herd of cape buffaloes, estimated at three hundred.

As we approached Mombasi the next day there were signs of some concentration of farming. In addition to fruits and vegetables there were fields of sisal, and some of it had been cut for drying so that it could be sent to Nairobi and made into ropes. There were also some petroleum refineries.

Mombasi is a resort town on the Indian Ocean, and we were ready for a little resort atmosphere. On a tour of the city we visited a museum of ancient relics, a fort built by the Portuguese in 1593, and a lot of shops aimed at tourists. We relaxed for a day, and walked along the shore of the ocean.

On the way back to Nairobi we had a stop at Voi Safari Lodge in Tsavo Park. We saw many interesting baobab trees and occasional animals as we rode along. Our hotel lobby had a very high ceiling, and probably thousands of bats hanging on it and sleeping. We did not go down to see them take off for their nighttime activities.

One incident during our tours to look for animals was a little bit frightening. The roof of the bus was open, and a baboon jumped up on top and was just about to join us. George speeded up quickly, and he went away. Always ready with my camera, I got a picture of his long arm practically touching Georgia.

We had time in Nairobi to shop a little at the market and at a craft shop. We left for a flight back to Brussels, and again had a holdover at Holiday Inn before leaving for New York and home.

ANIMALS AND BIRDS RECORDED IN KENYA

Animals
1. Elephant
2. Giraffe
3. Wildebeest (Gnu)
4. Cape Buffalo
5. Thomson's Gazelle
6. Impala
7. Grant's Gazelle
8. Bush Pig
9. Bush Buck

Birds
1. Ring-necked Dove
2. Kite
3. Ostrich
4. Tawny Eagle
5. Fish Eagle
6. Flamingo
7. Maribou Stork
8. Egret
9. Pelican

10. Water Buck	10. Cormorant
11. White-tail Mongoose	11. Secretary Bird
12. Brown Mongoose	12. Blue Heron
13. Striped Mongoose	13. Ibis
14. Sikes Monkey	14. Vulture
15. Red Duiker	15. Super-glory Starling
16. Tiny Squirrel	16. Weaver Bird
17. Jackal	17. Tick Bird
18. Hyena	
19. Topi	
20. Kapu	
21. Crocodile	
22. Hippopotamus	
23. Wart Hog	
24. Dik-dik	
25. Miniature Mongoose	
26. Rock Hyrax	
27. Eland	
28. Zebra	
29. Hartebeest	

Namibia, Botswana, Zimbabwe, South Africa, 1990

This trip was by TWA, and we had a day of waiting in Frankfurt. From there to Windhoek we were among the first passengers to fly in a Namibian Airlines plane. We had been scheduled for the South African line, but only a few days previously Namibia had become independent of South Africa. One advantage, we were told, was that our flight would be shorter. Namibian planes were allowed to fly over the continent, but South African ones, because of their racial situation, were banned from this airspace and had to take a longer route over the ocean.

In Windhoek we had a chance to rest and to have a tour of the city. Our guide reviewed the history of the country and mentioned some of the advantages and disadvantages of their independence. On the way to Etosha Game Preserve we passed through fairly flat land with scattered acacia trees and dry grass. There were a few cattle and sheep farms, and some areas where game animals were being raised for sale to zoos.

After we entered the Etosha Preserve we began to see large numbers of animals. Some which were new to us included three species of antelopes: oryx, kudus and springboks. The oryx were quite large, and had horns about three feet long, slightly curved backward. Their bodies were mostly white, with some irregular reddish-brown splotches. Kudus were also large, and had twisted horns. Springboks were probably more numerous than any other animal in the Etosha. They were small and dainty, and were recognizable by their habit of bounding suddenly into the air as if on springs.

A mother cheetah and her young one gave us our most exciting experience. Cheetahs are, of course, noted as the fastest land mammals, but this one was walking at a leisurely pace. The young one, looking very much like one of our kittens, was following her. He was in a playful mood, and would jump up, roll over, and investigate what was around him. The mother would stop now and then and wait until he caught up with her. We watched them with binoculars for a good while as they moved along what appeared to be a dry stream bed.

At one of the water holes we watched a series of animals as they came to drink. First there were kudus, then oryx; giraffes then gathered from all directions, followed by zebras and then by a determined-looking band of gnus. Each group left as the next one came. Another water hole was monopolized by a large number of elephants for as long as we watched.

Seeing the animals from a large bus did not allow us to get as near them as the little bus in Kenya, but we used our binoculars and saw many species. The impressive thing was the large numbers of several of the varieties. On the way back to Windhoek our driver-guide stopped to buy some kudu jerky and gave us a sample. It was hard to chew, but tasty.

We went next to Johannesburg, but only for one night and long enough to pack in a tote bag all we would need for the next five days. We then flew to Harare, capital of Zimbabwe. In my geography books this was Salisbury, capital of Rhodesia. Following several years of conflict and negotiation the country had become independent of England in 1980. There were still, however, a Victoria Museum and a park in which trees were planted in lines following the pattern of the British flag. We went to a tobacco market, said to be the largest in the world. The city gave the appearance of being rather prosperous.

After a day of sightseeing in Harare we flew to Lake Karibu, second largest man-made lake in the world, next to Lake Nasser. We transferred there to an eight-passenger plane for Buma Hills Lodge, on the other side of the lake. From there we went first to a sort of blind from which we watched animals come to a water hole. Then we had a boat ride around part of the lake shore, and a good many animals were sighted near the water. There was a spectacular sunset, and its afterglow on the water was amazing. At dinner that evening, the main item was ragout of impala.

The next day was our day to see Victoria Falls. The falls are about a mile wide and at one point are 420 feet high. This is about twice as high as Niagara, and one and a half times wider. The mist which rises from the torrents has been seen as much as thirty miles away. Because of this mist and the noise made by the water, the name the natives gave the falls is translated as "the smoke that thunders." There was a lane near the cliff facing the falls along which we could walk and get a continuous view. So much water spattered across on the trees, as well as on us, that it was like walking in the rain. We got soaking wet, but it was well worth it.

That evening we were entertained by a group of native dancers. Only the men were allowed to dance, but the women played unusual musical instruments and sang. The costumes, each different and created by the dancer himself, were the most outlandish get-ups imaginable. They used colorful cloth, beads, feathers, fur, and goodness knows what else. Each one also had his individual method of dancing. As the final number, the seventy-year-old leader did a remarkable belly dance.

The next morning we took the "Flight of the Angels" in small planes, flying over the falls as low as safety would allow. This gave a much better idea of the actual extent of the falls, and is another magnificent sight to remember.

After the flight we went to a sort of museum of former tribal life—a reconstructed village. The houses included a chieftain's dwelling, and there were tools and appliances which gave indication of life as it was carried on. One feature was several small huts on stilts. We were told that small children slept in these huts, and after they had been tucked into bed the ladders were removed until morning.

We had to give a report of the money we were carrying and fill out an entry form before crossing the border into Botswana. An unusual requirement was made because Botswana had learned that some of the cattle in Zimbabwe had "foot and mouth" disease. We had to get off the bus while it drove through a shallow tank containing disinfectant solution. Then we walked across the border on boards covered by a cloth which had been soaked in the same solution.

We went to Chobe Lodge, a lovely place situated on a lake, and our first activity was a boat ride. We saw several animals, and had an interesting view of a fish hawk catching fish. Another fascinating bird, the African jacana, was dancing daintily from one lily pad to another and picking up bugs or whatever he eats. On one edge of the lake there was a herd of Cape buffalo, so numerous that the heat from their bodies caused a cloud of mist to rise into the air.

On an early morning game-viewing trip the next day we had the unusual sight of a hippopotamus walking around out of the water. Then we watched as several black egrets spread their wings down low to make a shadow over the water and blind the fish so they could catch them. At least our guide said that was what was happening. There were many other interesting birds, including a spoonbill and a hatchet-top. Among the animals was a rare sable antelope.

Back at the hotel there were several wart hogs busily engaged in digging up the grass on the lawn to get the roots. That night wart hog medallions was one choice of meat on the dinner menu. It was very tough.

Next we went by small planes to a most unusual place called Xugana Lodge (pronounced "Chu-gana"), built on an island in the Okavango Delta. Yes, a delta in the middle of a continent is unusual. This one has been built because the Okavango River has its source in the high rainfall mountain area near the west coast, and flows down toward the interior as far as the edge of the Kalahari Desert. As it goes to lower elevations it branches in a fan-like pattern into smaller and smaller streams which eventually soak into the ground or evaporate. This pattern covers many square miles, and there are islands and lakes within it. Our planes deposited us at the edge of one of the branches. We transferred to a motorboat, and our ride to the lodge was through a maze of channels lined by a heavy growth of papyrus. We wondered how our boatman could possibly know which turns to take, but we finally arrived at the lodge.

Our housing was in large tents, similar to those at Kishwa Timbo, but with a bathroom for each two tents back between them. Electricity was turned off soon after dinner, so we were glad we had remembered to bring flashlights.

The next day we had a longer venture through the papyrus-lined maze to a lake where we saw a good many hippopotamuses. There were several other species of animals along the way, and several interesting birds. We had been told in one of the travel brochures that we might get a glimpse here of the "rare and elusive sitatunga." We did get a brief glimpse of one, and after I returned home I saw about a dozen of these animals at the North Carolina Zoo.

Even though this was a wonderful and unusual venture, it was somewhat of a relief to get to places which were more easily accessible. From a full-sized plane we could see the pattern of the delta and realized that with all of our travels through the channels we had seen only a tiny fraction of it.

We landed at Maun in time for lunch at the Duck Inn, said to be famous as a hangout for safari folk and other wanderers of the earth.

In Johannesburg we retrieved our big suitcases and got settled in our rooms. Johannesburg is a major industrial center known chiefly for its gold mining and processing. We spent most of our sightseeing time in the Gold Mine Park. We received there information about the geological processes which cause the gold-bearing minerals, through the ages, to settle in veins deep in the earth. While subsequent upheavals may bring it nearer the surface, the mines in this area were very deep. We were guided through a shop where some of the stages of processing gold were described and illustrated. Next we watched as an ingot of the final product was removed from an oven and poured into a mold. Then each of us in turn was allowed to lift a gold ingot (one which had been cooled, of course). It weighed twenty-eight pounds and was about half the size of one of our building bricks. Discoveries through the years have made it possible to retrieve a larger and larger percentage of gold from the ore but, in general, several tons of ore are required to produce one ounce of gold.

We donned hard hats and, carrying a battery-powered miner's lamp, went on an elevator deep down to a mining tunnel. As we rode along in trams, a lecturer explained the various processes and tools for digging and removing the ore. The many safety features were especially emphasized. It was truly an educational experience.

Rhinoceros—South Africa

The next day we had a flight to Sabi-Sabi, billed as a private and elite game preserve. The lodge was beautiful, the food was excellent, and we participated in such proper customs as afternoon tea and "sundowner" refreshments on the trails. For game viewing we traveled in Land Rovers, with a driver as well as a spotter who sat on the front of the hood with a gun across his lap. The drivers had walkie-talkies to communicate animal sightings to one another. Most of the animals were not different from those we had seen before, but an interesting variation was night viewing. The land rovers were equipped with floodlights, and we saw a number of animals that way.

The outstanding event of the Sabi-Sabi venture was that, after unfulfilled hopes and promises at several other places, we finally saw rhinoceroses. There was a group of about eight of them, and we watched them at fairly close range for a good while.

Back in Johannesburg for another night we got ready for a ride on the Blue Train to Cape Town. This very special train was quite handsome, the compartments were elegant and comfortable, and the food was delicious. We enjoyed looking at the beautiful scenery and later on had a restful sleep on the smoothest railroad tracks imaginable. Cape Town is a lovely city, crowded between the mountains and the ocean. We had a cable car ride to the top of Table Mountain for a magnificent view of the scene. The cable for this car, we were told, was the longest one without intermedial supports. It ran about two days out of every five, because of the frequent strong winds. We felt fortunate to be there on a calm day.

The next day we rode along the coast to the Cape of Good Hope. We readily agreed with everyone who had told us that this is one of the most beautiful places on earth. The view from the high cliffs, with the merging Atlantic and Indian Oceans beyond, was spectacular. As in several other times during my travels, it was hard to realize that I was really there.

We were invited to have dinner that evening in the home of a man who had some sort of interest in Olson Travel, the company in charge of our trip. Several friends of the family were also there, and we enjoyed meeting and talking with them.

The flight home from Cape Town, with changes and delays at Johannesburg, London and New York, was long and tiring.

Asia

✸ India, 1980

Our flight to India originated in Cairo and was a continuation of our trip to Egypt. We had a stop at Kuwait and a long wait in the gorgeous all-white marble airport. Back on board we realized that we had more passengers, and that the rule for putting hand luggage in the top compartment or under the seat had been suspended. We understood that these people from India, who had been working in the oil fields, were going home for a vacation. They were carrying television sets and other furniture and appliances. Then, if we had been amazed at that group, a stop in Bahrain brought in more of a similar group with even more unwieldy luggage.

In New Delhi we stayed in the elegant Taj Mahal Hotel. Our first sightseeing in the city took us to the handsome government buildings, built by the British during their occupation of India from 1917 to 1939. The British influence was evident in the buildings as well as in the uniforms of the guards.

During the tour, our guide, Mrs. Shah, told us that the major religions of India are Hinduism, Buddhism and Moslem. She explained that the basic belief of her religion, Hinduism, was that everyone must strive for perfection. Since this goal cannot be reached during one lifetime on earth, there must be incarnations. It may be necessary to experience several of these before perfection is achieved. Bodies of the dead are cremated, so that the soul can escape and enter another body. There is one god, who has no definite form, but is represented by three presences: Brahma the creator, Vishnu the preserver, and Shiva the destroyer. Each of them has several incarnations, many of which are represented by statues. The figure of Shiva has a slit in his forehead, which he can open into a third eye to look on evil and destroy it. Worship of cows is not a part of the religion, according to Mrs. Shah. Cows are useful because they produce milk and butter, and they are loved and petted, but "no one eats a pet."

We removed our shoes before going into a temple built for the worship of Vishnu. There were statues of Vishnu and his wife, as well as some of his incarnations. All sorts of other symbols decorated the interior, and it was very mysterious.

Again we had to take off our shoes to approach the memorial to Mahatma Gandhi, an enormous marble slab located in a garden with many flowers.

In nearby Delhi, the old capital, we went for a tour of the Red Fort, a large expanse of red sandstone buildings surrounded by a red wall. It was built by Shah Jahan in the early 1600s and was designed to shelter all of the city's inhabitants in case of danger. The king's palace and other buildings were well-preserved and very elegant.

Our most distinct impression of India as we moved from place to place was the great number of people. In the cities there were crowds everywhere, a few moving by taxis or buses, some on bicycles or rickshas, but most of them were walking. A few were dressed in western-style clothes, but most of the women wore saris, and the men wore loose pants and shirts or a dhoti (a long, loose wrap which served as trousers). Some moved as if they had a destination, but many were walking aimlessly. There were occasional stalls or stands where various articles were offered for sale, but the most memorable sight was beggars. Crippled or helpless old people were everywhere, and pathetic women with several sick and scrawny-looking children. Amid all of this, the big white cows sometimes wandered freely. We were warned firmly against giving money to the beggars. Once in a rural area we had stopped to watch women getting water from a well and carrying it in jugs on their heads. A group of children came up to the bus, and one of our group threw some coins out of the window to them. Almost instantly a mob of children and adults gathered and the driver had difficulty getting away without hurting some of them.

The next city on our schedule was Agra, site of the famous Taj Mahal. On the way we saw several animal farms—water buffalo, cows, goats, donkeys, or sheep. Most other farms seemed to be devoted to raising food for the animals, and women were seen carrying loads of fodder on their heads. Poor mud or brick huts often had piles of dung patties drying, to be used for fuel. Our lovely hotel in Agra was quite a contrast to the living quarters of the people.

I will agree with anyone who says that the Taj Mahal is the most beautiful building in the world. Pictures of it do not do it justice as it actually stands there in its majesty reflected in a pool surrounded by flowers. The white marble building of Moslem-type architecture is decorated by many carvings and by designs of set-in semi-precious stones. It stands on a platform 320 feet square and its central dome is eighty feet high and fifty feet in diameter. At each end there is a tall minaret. Thirty-six acres

Taj Mahal–Agra, India

of gardens surround it, and two identical mosques face it from its opposite ends. One mosque was built for worship and faces Mecca, as required; the other one was built simply to balance the functional one.

The Taj was built by Shah Jahan to serve as a mausoleum and memorial to his wife, who died giving birth to their fourteenth child. Their marriage was unusual for its time, in that it was a legendary love match. The building is said to have required the labor of twenty-five thousand people during a twenty-year period between 1629 and 1649. Undoubtedly many people are still engaged in caring for this and other show places which remain from the days of royalty.

During our entire visit in India we were served special meals which included meat, but once in Agra we decided to try the Indian vegetarian menu. Actually this meal was designed for us as "not spicy," but some of the dishes were pretty well seasoned. We remember some cauliflower in red sauce which burned our lips, not from heat, but from the hot spice.

During our rides in Agra and to the nearby city of Fatehpur Sikri, we were in a special bus driven by a Sikh.

He was so "upper crust" that he was separated from the passengers by a glass partition, and had a helper who opened the door for us and saw that we got on and off the bus safely. Our guide told us that the Sikhs are an offshoot of the Hindu religion. Young men in their teens are given their identifying turbans, which they are never seen without. We understand that they never cut their hair or shave their beards, and our specimen had a neat way of tucking his long hair and beard into his turban.

Fatehpur Sikri was built in the early 1400s by Akbar the Great, and is a combination of Moslem and Hindu architecture. Akbar had made a study of the various religions and had a dream of merging all of them. He designed a symbol which combined an arch for Moslem, a circle for Buddhism and Hinduism, and a cross for Catholicism. This design appeared on the columns of a number of the buildings. Another of his creative designs was a giant-sized parchesi board laid out in the pavement in front of his palace. Dancing girls stood in place on the various squares, and danced from one to the other as directed by his dice throws.

We had a brief flight from Agra to Jaipur. Our hotel rooms there had marble floors, but electricity was off from 7:00 A.M. to 7:00 P.M., and hot water, if we were lucky, ran from five to 9:00 P.M.

In Jaipur we visited the old part, known as the Pink City. It was built in 1727, and its pink walls and buildings still stand and are frequently painted. One especially interesting structure was the Wind Palace, with a large, artistically perforated screen in front of it. Behind it, we were told, the palace had a series of balconies on which the ladies could sit and watch the activities on the street without being seen. The palace had been converted in part to a museum, and another part was the residence of the current Rajah. Although he had no power or political influence, he was said to be the second-richest man in India. His source of income was mines which produce precious and semiprecious stones, and some quarries of marble, limestone and sandstone.

The original city of Jaipur was built on a nearby mountain as a fortress, but had to be abandoned because of lack of water. The old city had been partially restored, and we went as far up as we could by bus, then changed to elephant-back the rest  of the way. This was an interesting experience, and I had never fully realized before just how big an elephant is. The Indian elephants have long been used for transportation and as work animals, and respond very well to commands from the driver. The howdah carried four people, and our particular beast decided he wanted to be the first to get to the top of the mountain. What a ride that was! On the way we passed lots of people walking up the mountain to worship at the Hindu temple there. Other buildings included a luxurious palace called the Amber Palace. We walked back to the bus.

Elephant Ride—Jaipur

Our next stop was at Aurangabad to see the Ellora and Ajanti temples. Referred to as caves, the Ellora structures had been carved from solid basalt rock into the side of a hill. There were thirty-four temples in all; the first twelve were Buddhist, the next seventeen were Hindu, and the last five were Jainist. We were told that the temples were carved from the top down, and that the work was done by seven generations of artisans from A.D. 960 to around 1100. The work was unbelievably detailed, with minute designs in some places. The largest Hindu temple is thirty-six meters long, eighty meters wide and thirty-six meters high. It had many altars and symbols and figures of gods and animals. One impressive statue of Shiva showed him with eight arms; one was destroying an evil-spirited elephant, several were holding flowers, drums and so on and one was consoling his wife because he was away from her. I didn't figure that one out entirely.

We learned along the way that the Jains were formed from a combination of Hinduism and Buddhism. They believe that all living things, even plants, have souls, and they do not eat fruits or vegetables until after they have been picked long enough for the souls to leave them. They represent the ultimate in doing no harm to any living thing. Actually, with all that might be said in their favor, we thought that some of the figures in their temples bordered, at least, on the pornographic.

The next day we visited the Ajanta caves, with temples carved in the same manner as those of Ellora, but all of these were Buddhist. They are thought to have been carved between the first century B.C. and the ninth century A.D. After having been covered for centuries by jungle growth, they were rediscovered in 1819.

From Aurangabad we went to our last stop in India, Bombay. This city was much more like a Western city than any one we had visited before. Our attention was called, however, to a laundry with sheets, towels and clothing being scrubbed by hand and hung on lines to dry. The laundry man picks up and delivers the clothes. Our bus driver acknowledged that this was hard work and the pay is small, but asked how the laundry man would make a living if everyone had a washing machine.

On our last day a different kind of religious practice was noted at a place called the Temple of Silence in Malabar Hill above Bombay. This was the place at which the bodies of members of the Parsi religion are taken. This sect, originating in Persia and having about eighty-seven thousand members in Bombay, believes that air, fire, earth and water are sacred. Polluting any of these elements with dead bodies is forbidden. Their solution is to put the bodies on a platform on top of a hill and let the vultures devour them. With this gruesome thought, we left India the next day.

Tokyo, China, Hong Kong, 1981

We left home early on a Saturday morning and arrived in Tokyo Sunday night, having lost most of Sunday by crossing the International Date Line. The Imperial Hotel was beautiful and comfortable and the food was excellent.

We learned that Tokyo had eight million inhabitants— eleven million if the entire metropolitan area is included. We saw many buildings under construction, and our guide, Yatsu, told us that we would have to go to some of the older cities to see traditional Japan. She said that she, however, lived in traditional style, sleeping on a straw mat on the floor and sitting on the floor for meals. On our drive through the city we saw that several streets were lined by ginkgo trees. This goes back to the time when most of the houses were of wood and fires were frequent. Ginkgo is a "water tree," so they would stop fires from spreading from one block to another.

As we passed the government, or Diet, building, Yatsu told us that there were five hundred Diet members, only eleven of whom were women. However, she said that women were liberated and no longer had to walk three steps behind their husbands, but that most of them prefer to go "on a diet" rather than "to the Diet."

Religions are chiefly Shinto and Buddhism. Shinto has no Bible or other writings and its aim is to have a happy life. Prayers are made to objects in nature—trees, streams, mountains—and to national heroes and rulers. The place of worship is a rather simple shrine, usually with a Torii

Torii—Temple Gate, Japan

gate near the entrance. Buddhism is more formal, requiring much study of philosophy. There is worship of Buddha and of ancestors, and temples are elaborate. In one Buddhist temple we saw a little stand with something on it producing smoke. People were touching the smoke and then their bodies. This "Holy Smoke" is believed to cure diseases of the skin.

We had a short tour to Kamakura, which was the capital of Japan for a time after 1100. This is the site of a 750-year-old statue of Buddha, second largest and second oldest one in existence. It was most impressive.

From Kamakura we went next to Hakone for, first of all, a boat ride on Lake Hakone. Then we took an elevator to the top of Mount Hakone. A sign on the elevator read, "Capacity: 101 Japanese, 64 Americans." This was a ride which was to take us to see a good view of Mount Fuji, but we were disappointed because there was a heavy fog. Back in Hakone we admired an interesting outdoor sculpture museum before boarding the "bullet" train back to Tokyo. That was fast travel!

During our flight to Beijing the next day we had a great view of Mt. Fuji from the plane.

Our first sightseeing in China was a short trip from the Beijing Airport to the Temple of Heaven. This magnificent structure was built in 1400, and was used by the various emperors for their special prayers of blessings for the country. It is a cone-shaped building with three tiers, the bottom one of which is ninety feet in diameter. The cone-shaped roof is covered with blue tiles to blend with the sky. Wooden columns support the building, and no nails were used in its construction. The surrounding area has some interesting acoustic properties, including an Echo Wall.

Arrangements for the rest of our tour were in the hands of a young Chinese man whom we called by the un-Chinese name of William. Members of a youth organization had been assigned the responsibility for seeing to the details, and their leader had the more likely name of Chang.

On the way to our hotel we rode along a wide boulevard lined by newly-planted trees. The most impressive sight was the enormous number of bicycles. We were told that there were three and a half million bicycles in Beijing—one for each two inhabitants. There were practically no automobiles. A good bit of construction was going on, but there were also rows and rows of small shacks in which people were living. The most popular dress of men and women was a simple shirt and pants of olive drab or dark blue.

Our hotel, The Evergreen, was one of several buildings which had been purchased by the youth group to house their tourists. It had apparently been converted from a school dormitory. Our room had pink plaid bedspreads, ruffled pink pillows, and amazing hand-embroidered pink satin blanket covers. We were supplied with two large thermos bottles, one with drinking water and one with water for tea. Tea leaves and beautiful cups with lids were furnished, and the thermos kept the water boiling hot overnight and the next day.

The bathroom had everything it was supposed to have, but the shower nozzle was in the middle of the ceiling and there was no curtain. Also, all of our ingenuity was required to get one of the important items to work properly. The dingy appearance of the sheets and towels led Georgia to make the observation that all of the Chinese laundry men must have come to the United States.

The food was interesting and different. There were lots of dishes, some of which we could identify, but the rice was always reliable and filling.

We had been given a big buildup for a Peking duck dinner at the New Duck Restaurant. The dinner began with appetizers of liver and gizzards, and was followed by assorted vegetable dishes, some of which were quite tasty. The duck was then served with a flourish. It had been chopped into small pieces; after all, they do eat with chopsticks, but if our table of twelve had an entire duck it was very small and scrawny. My piece was a meatless wing tip. A special last piece was then brought in and placed at the center of the table. It was the head! To the consternation of the others at the table, I ate the brain.

Of course one of the special reasons for coming to China is to see the Great Wall. Construction of this wall began in the fifth century B.C. as a barrier against invaders. Building continued at intervals during the next

Great Wall of China—Tien Shan Mountains

several centuries. Its total length was about three thousand miles. The wall was made wide enough to allow six horses to run side by side. Signal towers were built about every four hundred fifty feet. The segment we visited is in the Tien Shan mountains, and we climbed uphill to the fourth tower. It was an amazing experience.

On the way to and from the Wall we passed several farming communes with rice, wheat, onions, cabbage and eggplant as the typical crops. There were no signs of farm machinery, and products were being hauled in horse-drawn wagons or small, often three-wheeled, trucks.

Another sightseeing tour from Beijing took us to the Summer Palace on Kunming Lake. This site is especially noted for a long covered walkway decorated with exquisite paintings. In the lake nearby we saw the Marble Barge, a ship built at the whim of an empress who apparently did not know that her elegantly designed craft would not float.

Near the center of Beijing is the Forbidden City, so named because it was the domain of the emperors and common citizens were not allowed to enter. The area of the Imperial Palace, built mostly during the early 1400s, covers two hundred fifty acres and has over nine thousand rooms. We saw the Treasure House of the emperors, with jewels, jade sculptures, ivory, carpets, and much more. The People's Republic of China had made the area into a museum open to all. It is said to be the most complete complex of ancient buildings which has survived.

Adjacent to the Forbidden City, and representing a striking contrast to this area reminiscent of ancient times, is the modern center of the People's Republic, Tien an Men Square. This square, covering ninety-eight acres, consists of a large open area in the center of which stands an enormous Monument to the People's Heroes. At one side of the square there is a large grandstand for viewing parades and celebrations. On another side is the Great Hall of the People, which can accommodate a meeting of ten thousand delegates. Five thousand guests were present there in 1972 for a dinner attended by Richard Nixon. Other buildings include the Museum of the Revolution, the Museum of History, the Cultural Palace of the Minorities, and the Memorial to Mao Tse Tung. I don't know what changes have taken place since 1981, but we viewed the body of Mao in his crystal tomb, and an enormous portrait of him covered a portion of the wall surrounding the square.

While in Beijing we had several opportunities for shopping, visited a zoo to see pandas, and were entertained at a variety show featuring acrobats, magicians, and so on. Wherever we went the people seemed glad to see us. While the old custom of binding the feet of baby girls had been renounced years ago, we did see two old women who had been victims of that practice. Little children often wore pants with a wide split in the seat, apparently made to eliminate the use of diapers. We were not sure how that relates to sanitation.

We left Beijing by plane for Xian (or Sian)—about a two-hour flight. This city was prominent in the early history of China, having been the capital from 212 B.C. to the beginning of the tenth century A.D. under the T'ang Dynasty. Its favorable location on Asian trade routes gave it wealth and prominence. After the fall of that dynasty, the city's influence decreased and it became somewhat isolated until a railroad connected it to other cities in 1936. Since that time it has become a center for manufacturing textiles and products useful in farming. Ruins of some of the ancient palaces and pagodas are still evident, and some have been restored. We visited a museum in which relics of the past were displayed. Of special interest were articles discovered by excavating a Neolithic village nearby. We then went to the actual site of the excavations and got an idea of the lifestyle of eight thousand years ago. One memorable site had apparently been a burial ground, and several clay jars contained bones which were obviously those of infants.

One of the attractions of the area has always been the presence of a hot spring about fifteen miles from Xian. A palace was built near the spring in the eighth century B.C. and other buildings were added over time. In 1949 the area was rebuilt and remodeled to make it a public resort. There are hotels, cabins, tea rooms and dining rooms, as well as picnic areas, and beautiful trees and flowers around a lake. Traditional Chinese architecture was used, with fancy roofs with dragons, lots of colorful designs, and a lot of red paint. We were delighted to have this sort of accommodations, after several days in the Evergreen.

The hotel complex was named Xianhuashengkai (pronounced "Sheean washing guy" and meaning "Xian watching the lake"). This name was embroidered on our pillows. The outstanding attraction in our quarters was a large bathroom with a sunken tub. This tub was about seven feet long, four feet wide and three feet deep. There was a raised step at the back to help one get in and out. The entire thing was lined with small colorful tiles. We were warned that the water came directly from the hot spring, with a temperature of 104 degrees Fahrenheit, so we had to let the water cool somewhat before taking a bath.

Of even greater attraction than the resort area was the archeological digging taking place not far away. The tomb of Qin Shi-Huag, a T'ang emperor who died in 210 B.C., is nearby. During his reign he was responsible for getting the various segments of the Great Wall connected so that it became a continuous boundary. He also made elaborate plans for his burial. A large mound in the area is thought to be his actual tomb, but there are also buried around the tomb ceramic figures of men about six feet tall, as well as statues of horses. Some of the men face the tomb, and others face outward as if to watch for intruders. The costumes of the figures represented levels of military rank, as well as uniforms of various regions of the emperor's realm. Some are kneeling, some are leading horses, and many carry weapons such as arrows or spears. Bronze was used for the weapons and for the reins of the horses. All of this was apparently in an underground vault and through the centuries the increasing depth of soil on the roof had caused it to collapse.

This underground army was discovered in 1973 by someone who was digging a well. Excavation began and a roofed building was constructed over a large section. We were allowed to walk on a passage around the working area and watch the people who were carefully extracting pieces from the soil. Occasionally there was a sizable piece, but some were very tiny, and every piece had to be meticulously brushed and labeled. One section of the building had been made into a museum to display the figures which had been assembled. There were also diagrams of the area being excavated, as well as of other sections thought to exist. Getting to view all of this was a most unusual and interesting experience.

Our transportation from Xian was by sleeper train car. In our capacity as first-class passengers we waited in a lovely room with blue sofas and chairs with lace antimacassars. Georgia and I were assigned a compartment with two other women of our group, who immediately claimed the lower berths. Georgia took an upper one, but due to the high temperature in the place and my aversion to upper berths, I spent most of the time in a jump seat in the cooler passageway. We reached Luoyang at about 3:00 A.M. and went to the International Hotel for the rest of the night. It was very plain, but had all the necessities.

Next morning we had breakfast of granola bars (brought from home) and some apricots saved from lunch the day before. Then we went for a walk but didn't see anything very interesting, so we came back and waited for an early lunch. Our sightseeing tour in the afternoon took us first to the tomb of General Guan Ya. There were four types of tombs in China, and this general and Confucius are the only two who ever rated Type One. The one we saw is only one of four tombs commemorating this man. Just his head was buried there; his body is in another location, his clothing in another and his soul is at rest in a tomb in his hometown. The one we saw was a good-sized and very elaborate building, with a large and fierce-looking image of the general in a fancy uniform.

Next we drove to a commune in which tractors were being manufactured. Twenty thousand people were employed in a factory which produced thirty thousand tractors each year, obviously for export since we never saw one in use in China. Families of the workers have their own housing.

Our next stop was at the Lungman Caves. This is a beautiful area with cliffs on either side of the Wei River, and lots of caves in the cliffs. There are carvings in many of the caves as well as in the rocky cliffs themselves. Our destination was a cave in which there was a forty-five-foot-high statue of Buddha. Around the large one there were many smaller ones, and others were carved in the walls. In fact, we were told that there are ten thousand figures of Buddha carved in the walls of that cave.

Back at the hotel after dinner we had a lecture by one of our tour group, Dr. Cooley, professor of Asian history at Guilford College. I should mention that he gave several talks at intervals along the way, making quite a contribution to our understanding of some of the history and customs of the Chinese people.

We had to get up at three o'clock the next morning in order to get to our train car, which had been sidetracked during our stop at Luoyang. Our daytime ride gave us an opportunity to see the countryside. At first along the river the land was flat, but we soon came to hilly and badly eroded areas. Wheat had been planted on almost any small patch which was reasonably level. It was harvest time, and people were cutting the wheat with a sickle and binding it into sheaves. As we came to flatter land, cabbage was being grown in irrigated patches, and there was still a lot of wheat. Water buffaloes were being used to draw primitive carts of sheaves to the winnowing places.

For our meals on the train we had to wait until the train had a stop long enough for us to get out and walk past nine cars to the dining car. After breakfast and dinner we again waited for a long stop and walked back. After lunch, however, there was no stop long enough so we walked back through the cars. They were very hot, smelly and crowded, and we felt that we were really being given a once-over. If you ever have to ride a train in China, do go first class.

During the afternoon we passed through more prosperous farming areas with a variety of vegetables. As we reached flat lands there were many rice fields, some of which were being plowed by water buffaloes. Before we reached our final stop at Shanghai we went through thickly settled areas, usually villages with walls around them. Then we had beautiful views of the Yangtze River.

In Shanghai we were in the Cheza Chiang Hotel, where Nixon had stayed. We were in the new south wing, and we suspected that Nixon was in the old part, since at that time in that country, old was best.

After breakfast the next morning a so-called cleaning team came to our room, carrying a dust cloth and a dirty mop in a bucket of black and greasy-looking water. They fanned the dust cloth over some of the furniture, gave the center of the room a few swipes with the mop, and were gone.

Our tour of Shanghai began with a look at the Exposition Hall, which had been built by the Russians. There were lots of displays of heavy industry and communication products, as well as large exhibitions and sales of arts and crafts. We could have completed our Christmas shopping there, but we had to go to see the Jade Buddha Temple and several pagodas.

Every large city has a Children's Palace, where talented children are taken after regular school hours for special training. There are also comparable facilities for older youths. In the Shanghai Children's Palace we were entertained by ballet, dramatic skits, and performances on several musical instruments. We also saw a science hall where some children were working with television and radios, and another group was involved with some sort of chemical experiment.

As a farewell to us, a group of children came and sang several songs. Then they indicated that they wanted us to sing for them. Someone in our group started "America the Beautiful" and others joined in, each in a different key, I think. It was so awful that the children giggled, and it was somewhat embarrassing.

We had an opportunity to shop on our own in the Number One Department Store in Shanghai. This was quite different from the Friendship Stores which are operated for tourists only. Actually it was more interesting to watch the people than it was to shop, but I did buy some pretty tea cups with lids.

We also had a visit to a rug-making factory. These were handmade rugs with a variety of patterns and colors. The wool threads were inserted into the basic warp and tied by hand. Some of the women were so adept at this that we couldn't actually see the knot being tied.

In a jade factory we heard a lecture on the various colors of jade and their sources, and then watched some carving. We were told that an intricate piece in the hands of one of the artisans would require a year for completion. The designs were mostly standard patterns, but each one is individually carved. Understandably, the prices of the finished pieces were very high.

Our next city was Hangchou, which we reached by train. As we rode along we again saw lots of rice fields, and a crop called oil seed which yields oil for cooking. Our hotel was located on West Lake, in a beautiful resort area.

A boat ride on the lake took us to the "Three Pools that Mirror the Moon" islands, and we admired the beautiful water lilies along the way. There was also a sort of botanical garden with trees and shrubs labeled with the sometime familiar Latin names. Back ashore there was another garden which featured hundreds of bonsai trees. A great many different kinds of trees had been trained to grow in the bonsai style, and we were told that some of them were about two hundred years old.

Again we saw Buddhist temples, and one enormous statue of Buddha carved from camphor wood. Another statue showed Buddha sitting on a lotus blossom, which is the symbol of peace and meditation. The prize-winning one, in my estimation, was Buddha sitting on a fish and holding a bottle which he is said to have emptied to form the oceans.

Another fabulous story was told to us at Yellow Dragon Cave. It began, "Once upon a time, long ago, even longer than that, the dragons who usually created rain by spitting out water went on a strike." Then it went on that one dragon refused to strike, since he did not want the earth to dry up, and the head dragon decreed that he was to be covered by rocks and soil, but he still spits out water. In the cave his head protrudes from the rocks and water flows from his mouth. It really does look something like a dragon's head might look, and it is yellow.

On a more down-to-earth level, we went to a tea commune. A terraced hillside was covered with tea bushes. They would grow to tree size, but are kept pruned for convenience in picking the leaves. A bush is cut to the ground after twenty years of production, but it will grow back in two years. We had been promised that we would see tea being picked, and although it was raining one of the girls took an umbrella and showed us the procedure. We also saw how the tea was dried in an electric apparatus, and heard about the different varieties of tea.

The tea commune is similar to others in that it has its own housing, schools, medical facilities, fields for growing part of its food, and recreation facilities. We visited the kindergarten and someone in our group had brought balloons for us to blow up and give the children, much to their delight. Although the commune is required to market its products through the government agencies, there is a possibility of getting a bonus for production in excess of a stated quota. This particular commune had used some of the funds secured in this way to purchase equipment for making silk brocade fabric. Proceeds from the sale of this material make extra income for the commune members. We watched workers preparing the coded cards which direct the loom to follow the specified pattern, and we also observed the weaving. Then there was a sales room where we could buy some of the products.

From Hangchou we had a flight to Canton. We had been on short flights on local airlines previously, but had not noticed any special difference from the usual type of procedure. This time was similar to others in that we did not hear any announcement of boarding time, but this plane was in a hurry. William gave us the signal to get on board, and the plane took off before all of us even got to our seats. There was nothing about seat belts, and we noticed that the plane had holes in the floor through which some of the machinery could be glimpsed. We were glad that this was not a long flight.

In Canton we were in the thirty-three-story-high White Cloud Hotel. It looked somewhat better than some hotels we had been in, but the floors could have used a good scrubbing. Then we noticed that we had a ragged window curtain, and realized that a window had been broken. The hole in the large window had somewhat the shape of a human body. We were on floor number 12-B (really number 13), and we wondered if someone had wanted to get rid of somebody. We hoped, however, that they had just wanted to get some fresh air, as it served that purpose for us.

The stop in Canton was only an overnight break on our way to Hong Kong, and we left the next morning by train. There were more rice paddies, and it also was evident that we were getting into warmer climates, because we saw banana, poinciana and frangipani trees. Rolls of barbed wire lined the border between China and Hong Kong. William told us that many people in China had attempted to cross the border, but very few succeeded, as there were strict guards with guns.

The Mandarin may not be the most beautiful, clean and comfortable hotel in the world, but it would have gotten my vote that night. It was like a different world, with carpets on the floor, comfortable chairs to sit in, really smooth white sheets, and a shining clean bathroom with soap, washcloths and fluffy towels. Then we had dinner in the top-floor restaurant and found out that clean tablecloths, exquisite china and glassware, and beautifully cooked and graciously served delicious food all still existed.

Our tour of Hong Kong started with a ride on the funicular to the top of Victoria Peak, where we had a great view of the large and crowded city of seven million people. Then we rode out of the city to a beach on the South China Sea. Next we went to a fishing village and saw the multitude of boats in the sea. Many people lived on these boats, because housing was not available on land. We were told that temporary housing had been built for over forty-five thousand refugees from China and Vietnam.

The biggest attraction for tourists is the enormous number and variety of shopping areas, in Hong Kong city and in Kowloon, a short distance by ferry across the bay. We had time allotted for shopping and took at least a limited advantage of it. When we realized how much we had accumulated during the entire trip, Georgia and I bought an expandable travel bag to use in getting it all home.

From Tokyo to Hong Kong, and all the places in China between, it was an outstanding and interesting trip.

Korea, 1982

In 1982 Georgia's son, Jim, was stationed in Korea, and his wife and children were with him. During May of that year, the United States and Korean governments made possible a visitation program for relatives of military members stationed in Korea. Of course Georgia wanted to go, and I was allowed to go along as her companion. The conducted sightseeing program continued for five days and we planned to spend about two weeks afterward with Jim and his family.

After greeting the family and getting a restful night in our hotel in Seoul we started our first excursion—to Panmunjom. On the way out of town we learned that Seoul had a population of eight and a half million. We were impressed by the great variety of carefully pruned trees and bushes along the streets. Mountains reached the edge of the city, and some houses were built on the slopes. In the countryside there were lots of rice fields, in various stages of maturity.

We had been given careful instructions as to how to behave in the Joint Security Area, and were briefed again after we arrived. We had to be dressed properly—no jeans or sneakers—and avoid looking directly or smiling at North Koreans or making any quick movement or gesture. With all of this in mind we were escorted to one of the buildings in which United Nations and North Korean representatives met for discussions. A table in the middle of the room had a marker down the center, in line with the exact border between the territories of the two participants. A North Korean flag and a United Nations flag stood close to the proper side of the line. We had been told a story about these flags. When the negotiations began, each delegation had brought its flag. One was slightly larger than the other, and at the next meeting the smaller had been replaced by a larger one. From then on there was a contest as to who would bring the larger flag, until they were so large that they wouldn't go through the doors. So, a compromise was reached. Each side would get a flag of reasonable size and place it on a stand. Then, if one stand was a little taller than the other, the shorter one could be made wider; and if one flag was a little wider than the other, the narrower one could have fringe. The results of this arrangement were very seriously pointed out to us.

We were allowed to walk around the table to the north side, and then were told about some of the problems which had been discussed there. Meanwhile a group of North Korean soldiers had gathered outside and were tapping on the windows and calling to us in an attempt to get our attention. To our credit, hard as it was, none of us turned our heads to look at them.

Later there were more stories of ways in which the North Koreans had practiced one-upmanship by making buildings on their side of the line taller, putting flags higher, and even making walls which they painted to look like buildings. We were also told about some of the really serious aspects of the situation, with constant threats of danger.

Back in Seoul we went for a visit to the Changduh Palace, one of four palaces of the Yin Dynasty in the city. We were told that the last living one of the princesses of the dynasty still lived in one of the buildings. The complex reminded us somewhat of the Forbidden City in Beijing, with beautiful oriental-type architecture and gardens.

The next day we went on a special train to Kyongju, the ancient capital city of the Shilla Dynasty. There were mountains and rough country along the way, but small villages were plentiful, and any level patch of ground was used for growing rice or ginseng. There were also vineyards and apple orchards. In Kyongju we had a tour of the historic area, including the tomb of King Michu, a third century ruler.

From Kyongju we went to the Ulson Industrial Complex, to see the enormous shipbuilding and automobile works. Development of the area had begun in 1972, and was owned in its entirety by Hyundai (a name which we saw on many cars and trucks in Korea). The complex was actually like a separate city, in that housing, shopping centers, schools and so on were provided for the workers and their families.

Back in Seoul the following day we had a trip to the Korean Folk Village, about an hour's bus ride from the city. This area has a series of homes with appropriate furnishings, representing different eras of Korean history and different classes of people. Types of houses ranged from thatched cottages to elegant buildings with oriental-type roofs. Each setup showed types of clothing, utensils, crafts and so on. We saw such activities as grinding rice flour, making paper from mulberry bark, weaving, painting, and making the unique celadon pottery. We sat on pads on the floor and had a Korean lunch which included bulgogi, noodles, bean sprouts, acorn jelly, kimchi, rice cakes, and several other things we didn't recognize. There was tea with the meal, but our dessert was Coca-Cola.

Jim and Susan came to take us to their house in Friendship Village. It was good to be where things were like home, especially the food. During the time we were there we enjoyed three special family occasions. First, there was Kerry's "Fly-up" in Girl Scouts, then J.P.'s graduation from kindergarten, and finally Jim's birthday.

Susan took Georgia and me to some truly great shopping areas. We bought bags of silk beads at one dollar a strand, eelskin purses and belts, and several other items special to Korea. We had gold pins made with our names in Hongul (the simple alphabet adopted to replace the numerous Chinese characters). We saw shops where unbelievable amounts and varieties of fabrics, flowers, baskets, toys and foods were displayed. Probably the most memorable food item was a large dog which had been dressed and cooked and was ready to be cut in desirable portions for customers. No, we did not buy any. Incidentally we were told that some American families which had brought their pet dogs with them to Korea had learned that this had not been a good idea.

We went back to the Folk Village with Jim and his family and had an opportunity to enjoy a great many fascinating things we had not had time to see on our other excursion. We also had a very interesting visit to a strawberry production area. This is so popular during the season that several entertainment features had been set up. There were merry-go-rounds and ping-pong tables, and other facilities. In the fields themselves there were tables, shaded by large umbrellas, where one could sit and enjoy a bowl of sweetened strawberries and a beverage. While many sodas were available, the most popular drink to accompany the berries seemed to be beer.

Jim and Susan showed us around several areas we had not seen before, and we had meals in Korean restaurants and were entertained by native dancers. Being there under those circumstances was truly a special privilege.

Bangkok, Singapore, Indonesia, Manila, 1987

We landed at Bangkok after a twenty-eight-hour trip, which started at Fayetteville with stops in Charlotte, Chicago, Seattle, and Tokyo. We were indeed glad to get settled in a beautiful hotel.

Of course the first and most important place for sightseeing was the
Grand Palace. This is a large square
with a number of elegant buildings, surrounded by a wall
which is more than a mile
around. In addition to the
actual palace and other
government buildings
and residences the temple of the Emerald

In the Palace Square—Bangkok, Thailand

Buddha is located there. The unique style of Siamese
architecture, used throughout the construction, is distinguished by elaborate tiered roofs and spires. The throne
room and other special areas are used only for state occasions, and a good bit of space had been converted to
museums and memorials. The temple of the Emerald
Buddha houses this revered figure and we were told that
large crowds of people come to worship on special days.

The royal residence is the site in which Mrs. Anna
taught the king's children, around a hundred years ago.
We were told that the movie made of that story may not
be shown in Thailand. The people consider that their king
was not portrayed with proper respect and dignity.

We were reminded that in contrast to other countries
in that part of the world, Thailand had never been ruled
by England, France or any other outside country, but had
always been independent. The present king is descended
from a long line of rulers and is highly respected.
Although he has no political power, he is said to be noted
for projects aimed at improving the life of the people.

We enjoyed a traditional Thai dinner, for which we
removed our shoes and then sat on pads on the floor.
Very ingeniously, there was a cutout space under the

table, so that it was quite comfortable. There were several courses—chicken, beef, and vegetables—each served separately with rice. Between courses, a few spoonfuls of soup had to be taken to clear the taste buds. A program of dances, depicting the history of the people, followed the dinner.

One of our excursions was a boat trip up the Chao Phraya River and into some of the many canals which branch from it. There were lots of houseboats on the water, and many small houses along the banks. We saw people washing clothes, food and children in the river water. One unusual feature was little boxes resembling bird houses at many of the homes. These were usually decorated with flowers and banners, some quite elaborately. We learned that they were spirit houses. A spirit lives on each piece of land, and when its home is destroyed by clearing the land and building something on it, that spirit must have a new place to live.

As we neared a market place we were met by boats with ladies selling flowers, fruits and so on. In the market itself we had an opportunity to see a wide variety of interesting craft products and to make purchases.

Back in the city we paid a visit to Thompson's Silk Shop. Spectacularly beautiful items were available. It was very tempting.

The time came for us to go to Singapore to board our cruise ship. First, however, we had a brief tour of the unique city of Singapore. Each major nationality which has settled there has its own center, and we went first to "Little India." Most of our time there was spent in walking around the food shops, of which there were many of great variety. Chinatown had one site of special interest. A house had been burned some time previously, and several people had died in the fire. A pomegranate tree had been planted in front of the house, and was decorated with red tags and ribbons to drive away the evil spirit which had caused the fire. No one would enter the burned house, or try to repair it.

Singapore has, of course, had a long history of colonization and capture, but has been a free nation since World War II. There is a president, but the government is controlled by a prime minister and a senate. It has strict laws against crimes, so crimes are few. Possession of certain drugs calls for a death penalty. The city is very clean, because of the severe punishments for littering.

We had a visit to the botanical gardens, but had time to see only the orchid area. It was unbelievable!

Our first stop on the cruise ship was at Semarang, on the Indonesian island of Java. The chief attraction was the temple of Borobudur, the largest Buddhist temple in the world, built in the ninth century A.D. It is four hundred feet square and one hundred five feet high. There are thousands of statues of Buddha, of a variety of sizes and postures. An earthquake and volcanic eruption in 1006 damaged the temple to the extent that it was abandoned to the elements and tropical vegetation until 1814. Some clearing and restoration took place then, and again in 1907, but the real reconstruction was done under the leadership of UNESCO from 1960 on. There were several tiers to the building, and one must go to the top level in order to achieve Buddhist perfection. We were there in February, and although their summer lasts all year, their hottest season is during our winter months. The temperature that day was 105 degrees, and the sun was almost directly overhead. After walking some distance in the shade of an umbrella which we had been told to bring, I decided to be satisfied with less than perfection and climbed to the second tier only.

Although the temple is Buddhist, my other outstanding memory of Semarang is related to Hinduism. This religion was brought to some of the Indonesian islands by Hindu princes who left India during the sixteenth century when Islamic armies came. The Hinduism they introduced merged in various ways with native worship of spirits and animals, and some strange practices emerged. We witnessed a ritual which was evidently an outgrowth of this union. A group of eight young girls, aged about ten years, performed a most unusual dance. They were dressed in fancy dresses and wearing elaborate headdresses. Each one was astride a board which had been cut in the shape of a horse and was gaudily painted and decorated. To the monotonous rhythm of a sort of band, they danced around in circles, carrying their "horses" with them. Just behind them were two young men smoking cigarettes and blowing the smoke toward them. (We were not told what was in the cigarettes.) Our guide told us that the object of the dance was to get the girls in a spiritual trance, and they did look as if they were mesmerized. It went on and on, and I don't know what would have happened if it had not started to rain.

Our next stop from the ship was at Bali, but we had a very short time there. Since my next trip to the South Pacific provided much more time to experience Bali, I will tell about it later.

We crossed the Java Sea, the Makassar Strait, the Sunda Sea, into the South China Sea, Manila Bay and Manila. After a brief tour of the city, we were entertained by a program which traced the history of the Philippines through dances of the various eras. It was very well done, with effective costuming, and program notes explaining each dance and giving a brief history of the period it represented.

We then had an opportunity to visit the American Cemetery and Memorial. This complex covers 150 acres and has graves of 17,206 members of the United States military forces killed in the Pacific area during World War II. Most of these had been moved to Manila from temporary graves in various other places, and represent about forty percent of those whose bodies were recovered. Others are in cemeteries on several other islands. The memorial, in the center of this large circle of graves, consists of a chapel, a tower, and two "hemicycles" enclosing a memorial court. Each hemicycle has twenty-four pairs of walls on which are inscribed, by states, the names and a bit of other information about the 36,279 Americans who were lost or buried at sea. There are also maps of the various areas in which wartime activities took place.

Georgia had another interest in going to Manila because she was born there. Her father, an Army doctor, had been stationed there at the time. The family moved to another location when she was too young to remember about it, and she had not been back since. She had been told that they lived at Fort McKinley, and when she asked about it she was told that the cemetery and memorial had been built on the site of that base.

As our ship moved out of Manila Bay, the "Rock" of Bataan was pointed out to us. We went on across the South China Sea and had a stop at Canton, with only enough time to visit the beautiful and impressive Sun Yat-sen memorial. From there we went to Hong Kong, for a brief tour and some shopping before our flight home.

✠ *Siberia, 1993*

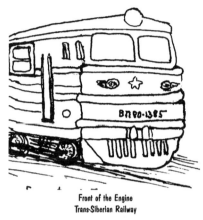

Front of the Engine
Trans-Siberian Railway

The chief attractions offered by this trip were the opportunities to see Lake Baikal and to ride the Trans-Siberian Railroad. Planned by the Smithsonian Institution, our flight took us to Anchorage, Alaska, and then across to Khabarovsk. After a short stay there we flew to Vladivostok and from there to Irkutsk, near Lake Baikal. We rode the train back to Khabarovsk.

Khabarovsk, located at the confluence of the Amur and Ussuri Rivers, was founded in 1895 as a center of trade for far-eastern Russia. It has grown into an industrial and political city of about forty-five hundred people. The Communist regime had ended almost two years prior to our visit, but our local guides did not comment upon the effects of the change, and we did not ask. The enormous Lenin Square with appropriate statue was still there, and the main street was Karl Marx Street.

Our flight to Vladivostok was delayed, reportedly by threat of a tidal wave, and we reached the airport after it had closed for the night. The lights were out, and in order to get to our bus we had to walk around the building through a treacherous area of all sorts of planks, bricks and other junk. There were no casualties, however, and we were taken to a comfortable, if not elegant, Intourist Hotel.

Vladivostok is in a strategic location near the tip of a peninsula which juts into the Sea of Japan. It is a center for shipping, and has a large naval base. For security reasons foreign visitors were not allowed there for a number of years, but this restriction was removed in 1990, and we were among the first tourist groups to be admitted. We had a tour of the city and were impressed by the large industrial port and the naval base. An old submarine which had been converted into a museum was included on the tour.

Irkutsk was founded in 1652 as a center for fur and tea trade routes to China and Mongolia. In addition to the traders who settled there, a number of exiles from European Russia came there. Often these exiles were intellectuals and members of the nobility who had displeased the ruling Tsar. Outstanding among this category was a group called the Decembrists who had, in December of 1825, organized a march protesting the Tsar's treatment of the peasants. They were sent there as prisoners, but gradually regained some freedom and were finally granted amnesty. They were not allowed to return to their homes, however, but their families were permitted to come and live with them. The wives brought home furnishings, art objects, books, and so on, and the city developed into a cultural center. We went into the former home of one of these families, which had been converted into a museum of the Decembrists.

When the railroad connected the city to Moscow and hence to other parts of Europe, materials to support industries could be brought in and the city became a center for manufacturing as well as for trading. Buildings in Irkutsk are much more beautiful than those we saw in any other Siberian city, and the entire area is attractively laid out.

A forty-mile bus ride took us to Lake Baikal, the world's deepest and second largest fresh water lake. It is over five thousand feet deep, four hundred miles long, and twenty to forty miles wide. The water is very clear and is safe to drink because of a type of sponge which grows in the water and filters out impurities. We were told that twenty percent of the world's fresh water is in this lake, and that if all the other drinking water on earth should run out, Lake Baikal could supply the entire population for forty years.

We spent the night in a hotel in the lakeside village of Listvyanka, and met some scientists involved in checking the lake to insure that it retains its purity. Among the sites in the village was a museum displaying many of the animals and plants which live in and around the lake. Some of these living things have never been found any other place on earth.

When we told friends that we were planning to go to Siberia they often asked us why we wanted to go to such a cold and barren place. Of course a large portion of Siberia is cold and forbidding, and thousands of criminals have been sent to forced labor in the Ural Mountain salt mines or in gold or coal mines. Our trip was in August, and we were in the southern part of the country, not very far from the border with China. The weather was moderate and we were comfortable. There are forests of coniferous trees scattered in the area, some parts which can be used for pastures, and even a few spots which can be used for growing crops. Even in that area, however, the winters are severe, and in December, Lake Baikal freezes to a depth of ten feet.

Stories about the construction of the Trans-Siberian Railway tell of many difficulties and setbacks, and give credit to the great determination and persistence which made the project possible. It was started in 1891, and segments of it began to take passengers in 1900. Our group was assigned three cars with twelve compartments each. We were given numbers to put on our suitcases and they were lined up in order so that the men with us could pass them, "bucket-brigade" style, to the proper compartment. In other words, stops were very brief, and there were no porters. Each compartment had two reasonably comfortable beds, each with a lamp, and a small table under the window between them. That was about it, except for some hooks on which to hang things. At one end of the car there was a toilet and a basin with cold water only. At the other end, there was a shower, but the little cubicle was so filled with all sorts of things that a shower was not possible. If one was lucky enough to be able to get there when the water was hot, there was a basin which could be used for a sort of bath, but it was not worth the trouble.

There was a dining car for our group, and the food was usually edible. Our Smithsonian planners had brought along some supplementary food, such as fruits, which helped. We were able to use the dining car at times between meals to hear lectures by our knowledgeable and interesting Smithsonian study leader. He was a professor of Russian language and history in one of our universities and had lived in several parts of Siberia in order to study the area firsthand.

We had left Irkutsk on a Tuesday morning and arrived in Khabarovsk late Thursday. A schedule of the many stops along the route was posted in the car's passageway.

At a few of these we were allowed to get off the train for five or ten minutes, but I did not take the risk of being left in one of those small towns. It was interesting to watch the activities at the stops, however.

The most pleasant and most unexpected sight was the beautiful scenery along the way. There were mountains, rivers, forests and open green areas. In addition to the pine and fir trees there were lots of stands of beautiful white birches. We saw gorgeous sunrises and sunsets. While looking out a train window at the scenery would not appeal to everyone, for me it compensated for any inconveniences along the way.

Back in Khabarovsk we had time to go one day to a museum of art, and to visit a most unusual shopping center. All sorts of merchandise, from fur coats to kitchen appliances, were on display. Most of this was outdoors, but there was a large building which had food supplies. Judging from the crowds of people who were buying things, we decided that at least the citizens of Khabarovsk were not suffering under the free enterprise system.

The next day we had a bus ride to a dacha on the Amur River. This beautiful building, in a lovely setting, had been used as a vacation spot for high-ranking communist officials. The people had taken it over as a recreation area and had prepared an attractive lunch for us. After the lunch there was a program of dancing and singing by a group of young people. The performance was good and some of the costumes were very elegant. It was refreshing to see Russian people who seemed to be having a good time.

We had a boat ride back to Kabarovsk, and I took a picture of the new Russian flag which was flying on the boat.

✖ Unusual South Pacific, 1993

When I saw the title and the itinerary of this trip I was intrigued, but had no idea how unusual some of the places could be. It began with Singapore, and we were offered a choice of a tour of the city or a cruise in the port area. Since Georgia and I had done the city tour some years previously, we chose the cruise. We had good views of the city, and the amount of activity at the docks was amazing. Many ships were waiting to be served. Our guide told us that Singapore ranks second only to Rotterdam in volume of shipping. We were also told along the way something of the rapid growth of the city. Things were so booming that there was no unemployment, and in fact attempts were being made to bring in workers from other countries to supply the demand for labor.

The other areas we visited are parts of Indonesia and Malaysia, and then there was Brunei. Because of the distances involved we depended a lot upon air travel, and schedules were sometimes quite complicated. For our series of visits in Indonesia we flew from Singapore to Jakarta. After a long wait in the airport we took a plane to Ujung Panang, capital of South Sulawesi. Following a brief tour of that city and a night there, we rode in minibuses to Rantepao. We were on flat land at first, then got into some beautiful mountain areas. There were several villages along the way, with houses usually built on stilts, and many rice fields.

As we neared Rantepao, home of the Torajas, the roads became very narrow and curved. People, goats, and chickens walked on the roads, so our progress was slow. Within the town the narrow streets were jammed with buses, trucks, and many bicycles. There were regular bicycles, but lots of them had a passenger seat in front of the driver. These, apparently, served as taxis.

Our hotel roof had a strange-looking projection that resembled a boat. We learned that this kind of roof was characteristic of the Torajan people. The story is that centuries ago they had come to this island from somewhere farther south, and had traveled up the rivers in boats. When the rivers

Home of Torajan Family
South Sulawesi, Indonesia

became too shallow for travel, the people settled down and built houses, using the boats for roofs. This custom has continued, with the boats always headed north, the direction of their progress. We saw several villages with rows of houses of this style.

The farther we got into the land of the Torajas the worse the roads became. Our first stop was to join a wedding party. Isaac, our English-speaking Torajan guide, assured us that we would be welcome, and indeed we were greeted warmly. The house was crowded with guests. The bride was sitting, with the groom standing beside her, at one end of the room. She was wearing a lavender dress, elaborately decorated with ruffles and lace, and a large headpiece to match. We greeted them as best we could, but did not get the impression that they were very happy. We were urged to stay for the wedding feast, but we were scheduled to go to a funeral.

Along the way Isaac gave us a review of the Torajan burial customs. A person is not actually dead until he is buried, because his spirit is still present. Sometimes several years pass before a deceased person can be buried, because a funeral requires the slaughter of water buffaloes. If the family members are not financially able to furnish a buffalo, they must make arrangements to get one, even if it means securing a young one and raising him to maturity. Friends may also agree to furnish a buffalo, and the more buffaloes there are the more prestigious is the funeral. These animals are sacred, and are not used as work animals.

Meanwhile the body of the deceased person is kept in the family room of the home on a sort of stretcher. A special "medicine man" comes and puts some secret material on the body and covers it. It then decays, but there is no odor. A bamboo pole from the stretcher goes through the floor and takes the fluids of decay to the earth. During this time the family members include the person in their conversations, speaking to him as if he were alive. When the time comes for the funeral, only the bones may remain to be buried. The spirit is buried with the bones, and the person is finally dead. The remains are placed in a crypt which has been cut in the side of a cliff. In some instances the place is marked by a picture or other memoir of the dead. Entire families may be buried in the same crypt, and we saw one of these which had life-sized models of the family grouped against the cliff.

Isaac had arranged for us to be guests at the funeral we attended, and one of a series of small buildings had been assigned to us. The slaughter of the buffaloes took place in a sort of arena, where crowds of people had gathered. The family of the deceased entered, dressed in black, and took their special seats in the reviewing stand. Seven buffaloes were slaughtered that day, but after watching the first two of them I decided to go to our designated guest house. I was greeted by two ladies who insisted upon serving me tea and cookies. I sat on the floor with my hostesses, and pretty soon some other members of our group joined us. One of the official buffalo slayers came around to greet us, proudly pointing to the blood stains on his clothes. After all, this was something sacred. Later on we understand that the buffaloes were dressed and cut up so that all the guests could take some of the meat home with them. We did not stay around for our share.

There were other strange things, including a large tree into which holes had been cut as tombs for babies. If they had died so young that they had no teeth and had not eaten solid food they were considered as not having lived. Their bodies should be given a chance to live by contributing to the life of something else. The tree showed several places which had grown over the tombs in varying degrees.

I did say that this trip was called "Unusual South Pacific," didn't I?

We went back to Ujung Pandang to get a flight to Denpassar in Bali. Our hotel, the Sheraton Laguna, was in the classy Dusa-Nua area.

About ninety percent of the people in Bali belong to the Hindu religion. They worship the same gods as the Hindus in India, cremate their dead, and don't eat beef, but they have additional gods and customs which are unique to them. At least one group lives by its own calendar, in which a year is made up of two hundred and ten days, or six months of thirty-five days each. They celebrate each New Year's Day, and we went to such a celebration at one of the oldest temples. The site by a sacred pool dates back to A.D. 900, but the original building burned and was replaced. Sacrifices of food were being brought to the gods of the temple, and it was amazing how high a stack of baskets of food the women could carry on their heads.

Next on our agenda was a performance of the traditional Barong and Keris dance. This was more of a dramatic performance than a dance, and represented the struggle between good and evil. A character dressed as a monkey was the good spirit, and a tiger was the evil one. We had a program explaining the various scenes, but it was hard to follow. All sorts of characters appeared, and there was lots of fighting and people changing into animals or witches, but we think that good triumphed in the end.

A few days later there was another performance called the "Kekak," or Monkey Dance. For this production there was a chorus of men representing monkeys sitting in concentric half-circles and chanting in rhythm, "Kekak-kekak-kekak-kak" over and over. The plot was another story of the struggle between good and evil, and the monkey again represented the good spirit. A princess was kidnapped, and there is much treachery and maneuvering, but she is finally rescued by the monkey general and his army. The continuous chanting all through the performance could be hypnotic.

For more down-to-earth activities we visited several craft shops. We saw expert and apprentice woodcarvers and looked at the products they had for sale. It was fascinating to watch the workers in a batik factory and to hear explanations of the various stages in production of a genuine batik piece. Beautiful fabrics were available. We also visited a jewelry factory.

Then came the big day! We were going to Komodo to see the dragons. The island of Komodo, too small to be represented on most maps, is a member of the Lesser Sunda Islands. The nearest airport is on Bima, another in that chain of islands. We flew to Bima to get a ship to Komodo. For this, we had been promised a nice ship with private cabins and other amenities. When we reached the dock, however, we learned that the only way to Komodo was on a small ship with meager accommodations. Our director took a look, then asked us to do likewise and decide whether or not we were willing to go. We saw that sleeping quarters on the lower deck consisted of a long row of double-decker bunks with no partitions. Since we had come so far and really wanted to see the dragons we decided to go. The only good things about the ship were the food, and the fact that we had a congenial group of sixteen people who were willing to make the best of a bad situation. There was a beautiful moonrise.

I was so tired and sleepy that I decided to try out the sleeping quarters early, and slept well until about three o'clock. Since I couldn't get back to sleep, I went to the bathroom and had a sort of basin bath. We had been furnished a flimsy towel, but I had the other necessities. The main problem was that the light would not come on, and my flashlight wouldn't burn unless my finger kept pressed to the button. But I made it somehow, and went back to my bunk and dressed in the dark. Most of the benches on the upper deck were occupied by someone sleeping, but I found a place, and there was a sunrise as beautiful as the moonrise had been the evening before. I was lucky to have done my bathing early, because I must have been the only one to have any hot water. In fact, only a few others got any water at all. After a good breakfast, however, we were ready for the dragons.

The Komodo dragon is a large, very large, member of the lizard family. It may grow to more than ten feet long.

Komodo Dragon

It is found native to no place on earth except Komodo and two smaller islands nearby. It looks slow and clumsy, but it can move quickly enough to catch a small deer. It also eats rats and other small animals, and has been known to attack people. The greater part of the island is a preserve where the dragons are protected as an endangered species. The rangers on the preserve must accompany any visitors, and there were several of them with us. We headed for a ranger station where the dragons often hang out. It was about a mile and a half from the coast, and the weather was very hot. We were not allowed to rest for a moment in the sparse shade of a tree, because a snake might drop out on us. When we got near the station, a dragon was walking across the trail right in front of us. He stopped and paused for pictures, and moved on. He was about eight feet long, but we saw larger ones at the station, and also saw several young ones. Incidentally, a few zoos have succeeded in keeping these animals, but they require very special treatment.

After we got back to our ship we enjoyed peace and quiet on our return to Bima. We went to a hotel which was described as "basic" in our literature. The rooms were large, with strange furnishings and plumbing, but we would have been all right except for the noise. Evidently the natives were Moslem, and when we heard a recorded chant we thought it was the evening call to prayer. However, it continued all night. The food was pretty awful, but we made it back to Bali in good shape.

After a lovely dinner and entertainment at the elegant home of one of the well-to-do local citizens, we had a restful night to prepare for our flight to Jakarta. Again we had a long wait at the airport, this time to begin our adventures in Malaysia.

The Shangri-la Hotel in Kuala Lumpur, capital of Malaysia, was billed as super deluxe. The city is very beautiful, with many new and modern-looking buildings and lovely parks and gardens. From Kuala Lumpur we flew to Kota Kinabulu, in Malaysian Borneo. Our tour of the city included various government buildings and museums, but the most remarkable sight was the thirty-story Sabah Foundation building. It looked something like a silo, and is unique because it has no columns supporting the floors. I didn't understand the method of construction as it was described to us, but it apparently is successful. It is an office building in which several organizations concentrate on research aimed at improving the health and general status of the people. We went to Kota Kinabulu because it is near the Sepilak Orangutan Sanctuary which we were to visit the next day.

A delightful walk through a virgin equatorial rain forest took us to the sanctuary. Most of the orangutans we saw were young ones which had been deserted by their mother or had been taken by someone who wanted to make the animal into a pet or to sell it. Extensive searches had been made to recover these animals, so that they could be brought back and trained to live in their native environment. They were fed and introduced to older members of their species, with the hope that they would become independent. Reports of success in this venture were encouraging.

We stood on a platform which provided a good view of the young ones as they swung along ropes which had been installed to bring them to the feeding station. This was a stand which encircled a tree about twenty feet from the ground. Men climbed a ladder to bring up milk and bananas. As the feeding time neared, we could see the animals coming on the ropes from all directions. It was fascinating to watch them swing along and to see them competing with one another to be the first at the food. One grabbed a large bunch of bananas and dropped it as he tried to get away with it. He came back for more. There was one large female who came with a baby clinging to her. We were told that she had probably adopted him. This was all quite an interesting show.

We went back to Kota Kinabulu and left early the next morning for Kuching. This city is also in Malaysian Borneo and is the capital of the state in which the famed headhunters of Borneo once roamed. It is a modern city now, but there are mementos of the headhunting days in the museums. In fact, one of our guides told us that his grandmother had belonged to a headhunting tribe.

From Kuching we went by bus to Ulu Skrang and took a longboat ride up the Skrang River to the home of the Iban tribes. The boat passed through rain forests, and occasionally there were areas cleared for rubber and pepper plantations. Deep in the forest was the home of the Ibans. We were scheduled to spend the night in the long house of that tribal settlement.

A long house is, as expected, a long building. It has a common passageway with a number of family quarters in the open area, with private rooms off to the side. We were required to remove our shoes before entering. Then we were shown an extension which was to be our quarters for the night. There were no furnishings or partitions— just a bare wooden floor. We began to have some misgivings, and these increased when we saw what was referred to in our literature as "basic modern facilities." This consisted of a flush toilet in a small cubicle off to the side of the building and some distance from our designated room. An unsteady bridge had to be crossed to get to it.

The life of the people there was interesting, and we saw something of their activities, their sources of food, and in general admired their ingenuity, but we began to indicate to one another that we were not happy with the overnight prospects. Our guide realized this, and asked us to vote on whether or not we wanted to go back to Kuching. The chances are that if we had not had the experience of the boat to Komodo we might have been willing to take this, but the vote indicated that we did not want to stay. This was a great disappointment to the local young tribesman who had directed us there and did not understand us at all. However, we enjoyed a beautiful ride back down the river and slept that night in the comfort and convenience of the Kuching Hilton.

For another unusual South Pacific venture we flew to Bandar Seri Bagewan, better known to us as Brunei. In this tiny country, a part of the island of Borneo, there are rich sources of oil, and the Sultan is thought to be the wealthiest man in the world. Evidently some of the wealth is shared with the people, for they seem to have nice homes and lovely surroundings. We saw the Sultan's palace, which we were told has seventeen hundred rooms, and his magnificent boat. There is a very large and beautiful mosque, which evidently was not up to his standards, and a larger and more eloquent one was being built. We spent a night in this unusual country, and next day we flew back to Singapore and home.

South America

❇ Peru, 1983

Our hotel near Lima, named El
Patio, was a cluster of adobe houses designed as a Peruvian village.
Senior citizens were assigned cottages near the dining room and we
were quite comfortable there. Our
tour company had arranged trips to
the city and there were buses for
us when we wanted to go on our
own. On the lawn in front of the com-

Llama—Peru

plex several llamas were doing the work of lawnmowers.
We noticed that each of them had a rope attached to a
collar, but the other end of the rope was not tied to anything. Evidently that gave them freedom to move around
the lawn, but kept them from wandering away.

The first tour into the city took us to the central
square, surrounded by the government palace, city hall,
other government buildings and a cathedral. We had a
tour of the oldest house in Lima, over four hundred years
old and the home of seventeen generations of the Aliaga
family. Members of the family were living in one wing and
the rest of the house was open for tourists. There were
many marvelous furnishings, including an ornately
carved mahogany bed, and architecture and ornamentation represented Spanish Baroque style. We also went to
a very old Dominican monastery, which was still being
occupied by members of the order.

Our next trip took us to Putuchucka, a site of Indian ruins dating prior to A.D. 1000. Evidently three different tribes occupied the area before it was conquered by the Incas about 1500. Digging and reconstruction began about 1902, and since that time a number of structures and artifacts of the ancient tribes have been discovered and partly restored. Farming must have been the chief activity of the tribes, as buildings and vessels evidently used for storage of grain were prominent. There was a very interesting stone device which was probably used for grinding corn. Tombs contained corn, beans and peanuts, along with clothing and jewelry. Pottery was made in the shapes of animals and vegetables.

During another visit to the city we went to a large museum of archeology, with displays arranged in sequence from prehistoric times through the Incas. Many exhibits were of pottery, and the improvement in techniques of pottery-making through the years could be traced. There were also large displays of gold articles, mostly jewelry, and a lot of mummies. An entire basement floor was devoted to pornographic pottery. Although we did not tarry long there, we got glimpses of some very explicit representations.

During our bus rides to the various places, our guide gave us background information about Peru. The chief exports are metals, especially copper, fertilizer, anchovies and coffee. He explained that we could not buy Peruvian coffee in Peru, since it was all sold to London and Paris before it was even picked. Government every now and then is by a military junta, but the most recent one of those was overthrown in 1980, and a president was elected. There was a high rate of unemployment, and strikes by the miners' union were in progress. School buildings were so scarce that three sessions per day were often required. There are usually separate sessions for girls and boys. Colleges and universities are so few that entrance is difficult. There was a teachers' organization which had succeeded in getting three days of sick leave with no necessity for reporting. According to our guide, all of the teachers get sick on the same three days.

It had never been known to rain in Lima. Occasionally there is a mist which may dampen things slightly, and it was estimated that this may account for one inch of water per year. There is plenty of water, however, because several rivers from the Andes run through the area. One residential section we passed had houses built of cardboard and other scraps of material and had no roofs. We could understand that there was no need for roofs as far as rain is concerned, but some shelter from the sun would seem desirable.

We drove down the Pan-American highway to Pachacamac, another area of archeological interest. On the way we saw sand dunes on which there was a series of large designs representing the emblems of several police units. The designs were made by plantings of bromeliads, a group of plants which require practically no water. Presumably the sand, being near the ocean, provided the needed moisture.

Pachacamac has ruins dating back to 200 B.C., and others up through the Incas. Earliest remains indicate it was a place of worship and ceremonial activities. There was a pyramid in the center where the Spaniards reported to have found a large emblem of the sun god, seven feet in diameter and four inches thick, made of solid gold. Of course the Spaniards took it back with them, along with any other gold they could find by digging into tombs and any other place they could find it. Restoration of the area began in the early 1900s, and was still far from complete.

During our stay in Lima we had meals in some interesting restaurants, often with entertainment. At one place some of the musicians used large and small boxes, which they beat upon or opened and closed rapidly to make the rhythm. One character played tunes on the jawbone of a horse. We also had opportunities to shop on several occasions. In addition to all sorts of ceramic and metal products, there was a great variety of rugs and other woven articles. I bought a beautiful shawl made of baby alpaca wool.

The alpaca is a member of the family to which llamas belong, along with vicuñas and guanacos. Vicuñas are found only in the wild, and are the smallest of the family. Their unusually fine fur has been so much in demand that it has become illegal to kill them. Alpacas are next in size. They are domesticated and are sheared like sheep. Llamas have been domesticated as beasts of burden, and are the most numerous of the group. Guanacos are the largest and are not domesticated.

A visit to Machu Picchu was actually the most compelling reason for the trip to Peru. In order to get to Machu Picchu we had to take a flight high into the Andes to Cuzco. This meant that we went from zero altitude to eleven thousand feet in about two hours. Walking from the plane to the terminal I felt as if I had lost several pounds. A bus took us to our hotel, where we were seated and given a cup of hot coca tea. This is the standard remedy there for altitude sickness, and we were strongly advised to lie down for a few hours after we drank it. The treatment must have worked, because we were ready for an afternoon tour which took us to a level of 12,200 feet for a look at the pre-Incan fortress of Sacsayhuaman. This structure was built of limestone, with each large stone cut and shaped in such a way as to allow a slight slippage in case of variation in temperature or earthquakes. It was a mystery that the people of that day knew how to shape the stones in this way. Another mystery was how these enormous stones, which were not natural to the area, got there in the first place. Considerable parts of the walls were still standing.

The first part of the train ride to Machu Picchu was a steep rise, and the train had to maneuver several switch-backs. Some of the mountains we could see had snow on their tops. Finally we came to the Urubamba River Valley, a branch of the Amazon, and rode along it for a good way. We reached our station in about four hours, but had to go the last lap on a rickety shuttle bus. The road was very bumpy and narrow, and it looked impossible that a bus coming from the other direction could pass. It was a wild ride!

The first view of Machu Picchu was amazing. There are remains of a city which was built upon a background of ter-raced and peaked mountains, and the buildings blend so perfectly with the set-ting that they appear to have grown there. Walkways and roofs fol-low the contours of the mountains. As we walked

Intiwantana—Machu Picchu

around we saw interiors of temples and houses, patches which had undoubtedly been used for gardens, large courtyards, and so on. One very interesting place was an observatory, with a kind of altar, which was called the Intiwantana, meaning "Hitching Post of the Sun." It is apparently a rock projection which was carved into a sort of throne so that the four sides are pointed to the four points of a compass. On its top there is a four-sided projection which could have been used as a sundial, since we were told that the slope of the sides is at the same angle as the latitude of the stone. Of course the slope is very slight, since the place is not far from the equator.

Back in Cuzco we had an informative tour of the city. It is thought to have been the home of several Indian tribes before the Incas took it over about A.D. 1040. From there the Incan empire spread over a large area of the continent. In 1534 Francisco Pizzaro conquered and claimed it for Spain. There were several attempts by the Incan leaders to establish a new capital, but none of them lasted very long. Machu Picchu is thought to have been built between the eighth and tenth centuries by a tribe which wanted an inaccessible place to carry on their culture undisturbed. For some reason it was abandoned for a long time. Some historians believe that it was the last resort of the Incas, and that Pizzaro was never able to find it. It is sometimes called the "Lost City of the Incas."

Meanwhile, Cuzco is a prosperous city, with some buildings built on the foundations of those destroyed by the Spaniards. Most of the inhabitants are Indians, a great many of whom have some Spanish ancestry. In fact, we were told that they are prouder of their Spanish background than of their Indian relationships.

✠ *Brazil, Argentina, Paraguay,* November 1990

When I saw Victoria Falls in June of 1990, several people who had seen Iguassú were comparing the two falls. Iguassú was considered by most of them as the more spectacular. When I got home and found in my mail a brochure announcing a November trip which would include Iguassú, I decided that I must go. I couldn't get anyone to go with me, but I managed quite well, and was pleased to learn that a friend from the Retired School Personnel was on the trip.

First we had a delightful time visiting Buenos Aires, with a marvelous guide named Don who was very proud of his city and loved showing it to us. We had wonderful food, especially the steaks. Buenos Aires was founded in 1580 and has ten million inhabitants—one-third of the population of Argentina. The name of the city means "favorable winds," and once in 1920 they had a light snow.

We went first to a section called "Boca," near the harbor, where a group of early Italian settlers had worked loading and unloading at the docks. Their houses had been made of packing materials and surplus lumber, and were painted with all sorts of combinations of colors, depending upon what was left from painting the ships. A portion of the site has been preserved and is quite colorful. An art colony occupies most of the area now.

The harbor is located on the Rio de la Plata, which means "silver river," in Spanish. "Argentina" comes from the Latin word for silver. Aristotle Onassis got his start operating a canoe ferry across the la Plata to Uruguay.

Soccer is the national game of Argentina, and in the cathedral chapel there is a statue called *Christ of the*

Soccer Players, given by the team which won the international title in 1978. There is a large and impressive memorial to St. Martin, liberator of Argentina from the Spaniards, in one of the parks.

One of our trips in the city took us to the very interesting Recoleta Cemetery. This is not a place with graves as we usually see them but is six city blocks of stone mausoleums, built very close together. They are of various sizes and ornamentation, and some are large enough to live in. It is permissible to have bodies on the ground level, and to have an underground floor. When Eva Peron died the people wanted her to be buried in her family mausoleum rather than with her husband. Since all the spaces were already taken, special permission was given to place her body in a vault on a lower third level. Eva was greatly admired because of her help to poor people. The play *Evita* is considered controversial and may not be shown in the country.

The tango, according to Don, originated with the Italians in the Boca. Only men danced at first, then the "ladies of the evening" were included. They wore the split skirts and were tossed around. A famous Argentine singer introduced the dance to Paris, and it became a respectable thing for the higher classes. We had front seats at a tango show in which some of the performers

Tango Show—Buenos Aires

were billed as the most highly rated in the world. The dancing was skillful, and some of the numbers had a dramatic theme. In one of them a teacher was trying to teach a pupil to tango, and it was hilarious.

Don gave us a report of the Falkland Islands War from the standpoint of a citizen of Argentina. The country had a military governor who was being criticized for the economic problems of the country and needed a diversion. So, he sent poorly trained troops from warm Argentina to the cold Falklands to drive out the British, with whom they had shared peaceful relations for years. Margaret Thatcher needed to boost her popularity, so she sent the entire British navy. People in Argentina were given false reports that they were winning, but the truth finally came to them. They ousted the military governor and started a democracy modeled after the United States. Everything in the Falklands settled down much as it had been before, but this did not bring back the large number of soldiers who had been killed unnecessarily. Don concluded by saying that they all hate Margaret Thatcher, and that they also hate Ronald Reagan for supporting the British.

The national drink of Argentina is maté (pronounced "ma-te" not "matay") tea, made from the crushed leaves of Yerba maté, a tree of the holly family. A special cup, often made from a gourd, is filled about half full of the dried leaves, and the cup is filled with hot water. A special sipper, made of metal or wood, has a wide bottom perforated so that the tea can be sipped without getting the leaves. Don said that the taste has to be acquired, and that at first it tastes like tobacco juice. The leaves may be used over and over, apparently, because our bus driver had a cup of the tea and at every stop he went into the restaurant or whatever and got a refill of hot water.

One day we went to a ranch in the country, driving down the Avenue of the Americas. On the way Don demonstrated the outfit worn by a gaucho. The pants are long and loose, with fitted cuffs so they won't be pushed up by the boots. A long woven belt goes around the waist several times, and is tied. Then there is a decorative leather belt with the emblem of a horse on the buckle. He carries three heavy rocks, covered with leather and each fastened to a long cord. These are swung and thrown so that they wrap around the legs of the cattle. A handkerchief around the neck and a poncho come next, and a knife in a fancy case fastens to the belt in the back. A flat-brimmed, black felt hat tops all of this.

At the ranch we were heartily greeted by the owner, who was wearing the gaucho outfit. He demonstrated some procedures for training horses, and some of our group members accepted his invitation to take a ride. Our meal began with large impanadas (spiced meat pies) and then we sampled the various cuts of meat. Dinner, served on tables under the trees, consisted of sausage, cole slaw, salad, and wonderful steaks. Dessert was a great variety of fruits. Another gaucho then came up, riding an enormous bull, and there was music and entertainment.

We were invited into the house to see some family pictures and some interesting furniture. A large display of maté cups, some fancily decorated with silver, were offered for sale.

After this venture we flew to Brazil in the neighborhood of Iguassú Falls. Right away we went back into Argentina to see the falls from that side. My first impression was amazement at the unbelievable amount of water and that it was colored by red mud. The rivers which make up the falls have their origin in the rain forest, and when first discovered and for some time thereafter the water was clear. Recent cutting and clearing in the forest has unfortunately brought on erosion.

We rode to several lookout points to get views, and walked on a kilometer-long catwalk over the river to Devil's Throat, a point at which the water makes an enormous whirlpool. Actually the spray was so heavy that we couldn't get very near, but we got a good impression of the force of the water. As we were going back to our bus we saw lots of interesting butterflies and two iguanas about three feet long.

From our hotel in Brazil we went the next day to see the falls from that side. There were several stops for viewing the beautiful sites, but it is not possible to see the entire thing from the ground. So, we had a helicopter ride to get the whole picture, including Devil's Throat. We then had a leisurely stroll along a path which gave a different view. On the way we saw a coati-mundi.

My comparison of Iguassú and Victoria Falls is that Victoria is amazing, clear and undoubtedly the more beautiful, but for majesty and power, and also beautiful views, Iguassú is the greatest.

Our tour director thought we ought to go to Paraguay, since it was so near, and we went to a little town just beyond the border. Since it is such a poor country and does not have to pay import taxes, Paraguay can get products from all over the world and sell them at low prices. There were unbelievable numbers of name brand watches, and myriads of displays of almost anything imaginable. Even so, I did not see anything I really needed or wanted.

The last stop on our agenda was Rio de Janeiro. I felt very special that I was given a beautiful room in the Sheraton Rio with a balcony facing the ocean. A tour of the city gave us views of the Sugar Loaf, the statue of *Christ the Redeemer*, and several parks and government

The Sugar Loaf—Rio de Janiero

buildings. With the ocean as its setting, Rio is justifiably called one of the most beautiful cities in the world. We drove to Copacabana Beach, and admired its distinctively tiled walkways and its lovely hotels. Our guide assured us that it was nothing but sand until someone took a chance and built the Princess Hotel. Helping greatly to make it famous was a visit back in the twenties from the Prince of Wales, later to serve for a short while as Edward VIII of England. The newspapers reported that the Prince went swimming "almost in the nude." Actually we were told that he wore knee-length shorts and a T-shirt.

One of the unusual buildings into which we went was a Catholic church. It was 290 feet high and shaped like a pyramid. The theory of its builders was that a series of louvers on each panel would create automatic air-conditioning. I don't know whether or not it worked. Almost all of the residents of Rio, according to our guide, register as Catholics, but many of them rely on the church only for hatching (baptism), matching (marrying) and dispatching (burial).

Brazil claims to have deposits of practically every precious and semiprecious stone. Amsterdam Sauers has a gem museum with many stones displayed in their natural state, and mockups of shaft and surface mines. There is also a room in which various gems are being cut, polished and placed in settings. Of course there was also an extensive sales area. It was a very interesting place.

We took a cable car to the top of Sugar Loaf and had a chance to walk around and look at the city and the harbor.

On the highest point of the city there is an enormous statue of Christ. We happened to be there when it was being given a new coating, and scaffolding around it pretty much obscured the view from the lower levels. However, we were able to take a cog railway up to it, and could see what it was like in spite of the scaffolds. It is truly magnificent.

We had a trip outside of the city into the Atlantic Mountains. There were beautiful views of tropical forests and waterfalls along the way. Our destination was Petropólis, a town named for the last emperor of Brazil, Peter II from Portugal. His palace is still there, and the town has become a resort and retirement area.

Our last night in Rio was spent at a Samba show. There was some singing and dancing and very loud music. Our guide had given each of us two small balls of cotton for our ears, and we gratefully used them. Most of the evening was a show of costumes, displayed by women on a walkway similar to the one used at Miss America contests. The outfits were extravagantly decorated with feathers and so on, and it got to be very boring. I guess I was tired and ready to go home.

Antarctica and the Falkland Islands, 1993

If you are interested in seeing something of the Antarctic, but don't care to combat blizzards, trek over ice with a backpack and sleep in tents, or if you are not qualified to participate in a scientific expedition, then go for a cruise. It is amazing how much you can see without going far from comfortable quarters and good service and food.

We flew from Miami to Punta Arenas, located on the Strait of Magellan on the Chilian portion of Tierra del Fuego. After a tour of the town we were welcomed to the *Ocean Princess* to begin our cruise. The next day was spent with lifeboat drills and getting acquainted with the ship in general. There were welcome parties, lectures, and a lot of time spent just looking at the magnificent scenery. At some point along the way we left the Strait of Magellan for the Beagle Channel and stopped for the night.

The next day we went ashore at Ushuaia, on the Argentine portion of Tierra del Fuego. It is the southernmost city in the world, and has a population of eleven thousand. Our tour of the city included an interesting botanical garden and a museum of machines and other reminders of the early settlers. Most of the business of the city is based on the woolen industry. We learned that except for the towns, the island population averages one human and forty sheep for each square mile.

Back on the ship during the next day we crossed Drake Passage and had more lectures on topics related to our anticipated excursions. Each of us was given a red parka along with instructions for proper dress and safety procedures. We were given colored tags—red, green, blue or yellow—which separated us into groups for going ashore. Each group was given a debarkation time, with the order rotated each day. We had capable help in the rather tricky procedure of getting into and off the zodiacs, and there were no problems. Of course it was cold, but we were dressed to cope with that, and we did not go on shore excursions when it was windy or when bad weather threatened. Each landing area had been scouted for safety beforehand, and our time ashore was limited to about an hour. Since we were there in January—their summer—the ice had melted for some distance back from the shore so we did not have to walk on anything more slippery than rocks. One member of the ship's crew had the sole responsibility of checking for the presence of icebergs, and we did see several of these. There were always spectacular views of ice-covered mountains in the background. We had some flurries, but no real snowfall. We had a landing almost every day for a week, and once or twice there were two per day. Most of these were on islands of the South Shetland group, but at one time we were on the Antarctic continent.

On every landing we saw multitudes of penguins, usually the Adele species. This was their breeding season, and many of them had young ones in the nest with them. The babies were fuzzy and their feathers were gray, in contrast to the black and white "uniform" of the adults. During most of the year they live in the water, and it was interesting to

Penguin & Chick
Antarctican Isles

watch some of them swimming and probably getting a
mouth full of krill to bring back to their young ones. We
were told not to get closer then fifteen feet to penguins
in order not to risk startling them. However, if we stood
still and a penguin walked nearer to us, he was breaking
the rule. This did happen now and then, as they were
quite friendly.

On a few occasions we saw elephant seals resting on
rocks near the shore, and once we saw fur seals. We were
told that after the whalers of the nineteenth and early
twentieth centuries rendered the population of whales
almost extinct, they started killing seals for their skin and
fat and almost annihilated them.

At one of our landings, Deception Island, we had an
unexpected experience. The island had been the site of a
volcanic eruption, and the ashes had not yet thoroughly
cooled. In some places the hot material had poured into
the sea and was still creating sufficient heat to cause
steam to rise from the water. It was actually warm
enough to permit swimming, and some of the bold ones
on our cruise had put their swimsuits on under their
warm clothes. They went near the water and undressed
down to their suits and ventured out. They did not go far,
however, because they found that the water got pretty
cold a short distance from the shore. Then they had to
put their other clothes on top of wet swimsuits. Safe
paths were outlined by ropes, so that we didn't walk in
places where the heat might have damaged our shoes.

Now and then we saw deserted stations which had
been used for research, but we never saw a sign of people
except for those connected with our ship.

Another day and a half at sea, headed north, took us
to West Point Island, a privately owned member of the

Falklands. We went ashore by zodiac, and were allowed to walk around and see more penguins and a variety of other birds. We were served tea in the home of the island's owners.

East Falkland, largest of the islands, is the site of the capital, Port Stanley. For our landing there tenders were used and we got ashore without having to take the last few steps in water. We walked around and looked at the government buildings and churches and the British-looking gardens. Handsome woolen items were for sale, at handsome prices. Some of the ladies who were selling them said that they had personally sheared the sheep, carded and spun the wool, and then knitted or woven the product.

The trip from the Falklands to Buenos Aires was uneventful for almost three days, but as we got near the harbor some very rough seas suddenly greeted us. We were going into the dining room for dinner when the ship lurched. We were banged around, but not hurt, and all of the dishes on the tables were flung to the floor. The waiters scrambled to get the tables reset, and we ate our dinner holding onto the table with one hand to keep our chairs from sliding. The ship had to anchor and wait for about twenty-four hours for the storm to subside, and then wait its turn to enter the harbor and dock. This delay cut short our visit to Buenos Aires, but we did get a short tour of the city. After another night on our ship, we said good-bye to the *Ocean Princess* and took a plane headed for home.

However interesting, exciting, and beautiful a trip had been, the best part of it was to see the gate to my driveway and home.

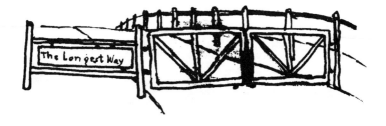

Appendix

Trips Outside the United States in Chronological Order:

1971:
    January 16-23, Caribbean Cruise (Puerto Rico, Haiti, St. Thomas, Bahamas)
    December 21-31, Russia (Leningrad and Moscow)

1972:
    February 16-20, Puerto Rico and St. Thomas
    June 11-July 1, Galapagos Islands and Ecuador
    August 7-15, Mexico

1973:
    May 10-25, Switzerland, Austria, Germany, Denmark, Italy
    December 22-January 1, Russia

1974:
    May 12-19, Jamaica
    August 11-16, Canada (Victoria and Vancouver)
    December 20-30, Mexico

1975:
    June 30-July 17, England and Scotland
    December 27-January 3, Caribbean Cruise (Jamaica, Grand Cayman, Cozumel)

1976:
    June 9-29, Denmark and Norway

1977:
    June 13-July 5, Austria, England, Wales, Scotland, Ireland

1978:
    June 21-July 6, Finland, Sweden, Denmark
    December 22-January 1, Spain

1979:
    October 30-December 1, Fiji, New Zealand, Australia, Tahiti

1980:

> February 26-March 17, Egypt and India
> June 12-July 1, Netherlands, Germany, France, England
> July 1-27, Canada (Ottawa, Toronto, Stratford)

1981:

> February 21-March 8, Brussels and Kenya
> May 16-June 4, Japan, China, Hong Kong
> October 11-25, Greece

1982:

> May 15-June 3, Korea
> September 9-23, Yugoslavia, Hungary, Austria
> November 5-23, Portugal, Spain, Tangier

1983:

> February 6-17, Peru
> May 20-June 12, Ireland

1984:

> All tours were in the United States

1985:

> January 19-30, Costa Rica, Panama Canal, Colombia, Aruba, Curacao
> June 13-29, Yugoslavia, Hungary, Czechoslovakia, Poland, East and
>   West Germany, Austria
> October 8-20, Sorrento and Rome
> December 22-26, Bermuda

1986:

> May 15-28, Caribbean Comet Cruise (Venezuela and many islands)
> July 2-15 Vancouver Expo and Canadian Rockies

1987

> February 4-22, Bangkok, Singapore, Java, Bali, Manila
> November 11-19, Garmisch-Partenkirchen

1988:

> November 13-20, Ireland

1989:

> March 27-April 3, Canterbury and London
> August 1-21, Greece and Greek Islands, Turkey, Israel, Egypt

1990:

May 24-June 15, Namibia, Botswana, Zimbabwe, South Africa

September 12-24, Switzerland

October 3-November 11, Argentina, Brazil, Paraguay

1991:

July 26-August 7, Canada (Nova Scotia, Prince Edward Island, New Brunswick)

August 11-21, Canada (Manitoba)

December 5-12, Mexico (Copper Canyon)

1992:

February 28-March 12, Colonial Mexico

August 8-18, Greenland, Iceland, Orkney Islands

1993:

January 23-February 11, Tierra del Fuego, Antarctic areas, Falkland Islands, Buenos Aires

August 8-23, Siberia (Khabarovsk, Vladivostok, Irkutsk, Lake Baikal, Trans-Siberian Railway)

October 30-November 22, Singapore, Sarawak, Bali, Komodo, Malaysia, Borneo, Brunei

1994:

May 27-June 5, Calgary and Canadian Rockies

1995:

June 20-July 1, France, Belgium, Netherlands, England